Make:

ReMaking History
Industrial Revolutionaries

Volume 2

Make:
ReMaking History
Industrial Revolutionaries

William Gurstelle

MAKER MEDIA
SAN FRANCISCO, CA

ReMaking History, Volume 2
Industrial Revolutionaries
By William Gurstelle

Printed in Canada.

Published by Maker Media, Inc.,
1160 Battery Street East, Suite 125
San Francisco, California 94111

Maker Media books may be purchased for educational, business, or sales promotional use. Online editions are also available for most titles (*safaribooksonline.com*). For more information, contact our corporate/institutional sales department: 800-998-9938 or *corporate@oreilly.com*.

Publisher: Roger Stewart
Editor: Roger Stewart
Copy Editor: Rebecca Rider, Happenstance Type-O-Rama
Proofreader: Elizabeth Welch, Happenstance Type-O-Rama
Interior Designer and Compositor: Maureen Forys, Happenstance Type-O-Rama
Illustration: Richard Sheppard, Happenstance Type-O-Rama
Cover Designer: Maureen Forys, Happenstance Type-O-Rama
Indexer: Valerie Perry, Happenstance Type-O-Rama

November 2016: First Edition

Revision History for the First Edition
2016-11-11: First Release

See *oreilly.com/catalog/errata.csp?isbn=9781680450668* for release details.

978-1-68045-066-8

Safari® Books Online
Safari Books Online is an on-demand digital library that delivers expert content in both book and video form from the world's leading authors in technology and business.

Technology professionals, software developers, web designers, and business and creative professionals use Safari Books Online as their primary resource for research, problem-solving, learning, and certification training.

Safari Books Online offers a range of plans and pricing for enterprise, government, education, and individuals. Members have access to thousands of books, training videos, and pre-publication manuscripts in one fully searchable database from publishers like O'Reilly Media, Prentice Hall Professional, Addison-Wesley Professional, Microsoft Press, Sams, Que, Peachpit Press, Focal Press, Cisco Press, John Wiley & Sons, Syngress, Morgan Kaufmann, IBM Redbooks, Packt, Adobe Press, FT Press, Apress, Manning, New Riders, McGraw-Hill, Jones & Bartlett, Course Technology, and hundreds more. For more information about Safari Books Online, please visit us online.

How to Contact Us
Please address comments and questions concerning this book to the publisher:

Maker Media, Inc.
1160 Battery Street East, Suite 125
San Francisco, CA 94111
877-306-6253 (in the United States or Canada)
707-639-1355 (international or local)

Maker Media, Inc. unites, inspires, informs, and entertains a growing community of resourceful people who undertake amazing projects in their backyards, basements, and garages. Maker Media, Inc. celebrates your right to tweak, hack, and bend any technology to your will. The Maker Media, Inc. audience continues to be a growing culture and community that believes in bettering ourselves, our environment, our educational system—our entire world. This is much more than an audience; it's a worldwide movement that Maker Media, Inc. is leading and we call it the Maker Movement.

For more information about Maker Media, Inc. visit us online:

- *Make:* magazine *makezine.com/magazine*
- Maker Faire *makerfaire.com*
- Makezine.com *makezine.com*
- Maker Shed *makershed.com*
- To comment or ask technical questions about this book, send email to *bookquestions@oreilly.com*.

Please visit *www.ReMakingHistory.info* for information including corrections, updates, and further reading recommendations.

Acknowledgments

Thanks to the crack editing and production team at Happenstance Type-O-Rama for their hard work.

Sincere appreciation to the staff at Maker Media for all they've done.

Thanks to my wife Karen for her invaluable help and support.

About the Author

William Gurstelle has been writing for *Make:* magazine pretty much since the beginning. Besides 40 or so "ReMaking History" columns, his work there has included do-it-yourself pieces on a gravity-powered catapult, a taffy-pulling machine, a Taser-powered potato cannon, and an ornithopter. He's also a bestselling author, a registered engineer, and a popular speaker on the world of science and technology.

Contents

The Dawn of the Machine Age

The projects in this book were inspired by inventions of the time period from about 1700 to 1875, a period often known as the Industrial Revolution (see Figure I.1). This was a very, very important time in the history of invention. Before this period, most things were made one item at a time by craftsmen like blacksmiths and stonecutters. But this was a slow and expensive way to make things, especially for items many people were interested in buying and using.

The 19th century saw the invention of many things we still use today; it was during the Industrial Revolution that great demand arose for a host of products like rubber, batteries, steel, and electric lights. Because we began to make these commodities in factories instead of workshops, we could indeed produce them in much greater quantities and at more affordable prices. This was the revolutionary idea at the core of the Industrial Revolution: we would use machines instead of human labor to make lots of stuff more cheaply, and doing so would be the way of the future.

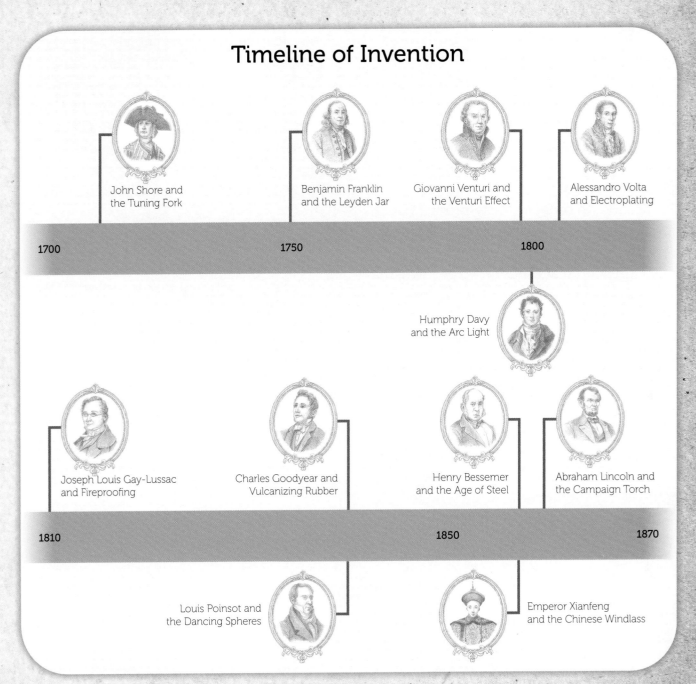

Timeline of Invention

John Shore and the Tuning Fork

Benjamin Franklin and the Leyden Jar

Giovanni Venturi and the Venturi Effect

Alessandro Volta and Electroplating

1700   1750   1800

Humphry Davy and the Arc Light

Joseph Louis Gay-Lussac and Fireproofing

Charles Goodyear and Vulcanizing Rubber

Henry Bessemer and the Age of Steel

Abraham Lincoln and the Campaign Torch

1810   1850   1870

Louis Poinsot and the Dancing Spheres

Emperor Xianfeng and the Chinese Windlass

Figure I.1: Timeline of Invention, the Industrial Revolution

The Industrial Revolution gave birth to a number of ideas and techniques that became the basis for today's modern world. Many if not most of the things you use in your daily life—from sneakers to salty snacks to smartphones—are produced quickly, efficiently, and relatively inexpensively in factories that use the techniques of mass production. In order to mass-produce our commodities, we use machines to quickly perform manufacturing operations, and although such machines are expensive, they can really chunk out a lot of products at a very fast pace.

Before the Industrial Revolution, people had to spend long, hard hours making just about everything by hand. This meant that nearly everything was expensive; cloth for clothing, flour for bread, and wood for tables and chairs were almost unimaginably pricey by current standards. But when machines came along and factories were built to house them, a lot more stuff became available, and it was far more affordable to obtain. The Industrial Revolution is the name we gave to the time period when machines changed the way people lived.

The Industrial Revolution began around the middle of the 18th century, when people in England began to build steam-powered machines to make cloth. Not long after this, steam power was applied to locomotives, and then mining equipment, and then just about everything that was previously powered by the muscles of humans and animals.

From England, the Industrial Revolution spread out to other countries in Europe and the Americas and then to the rest of the world. By the turn of the 19th century, it was nearly everywhere.

But the great changes of this era weren't due to the rise of steam-powered machines and factories alone. There were thousands, perhaps millions, of new ideas gushing forth that made for better and richer lives. And once a new idea became well understood, improvements to it came fast and furiously.

In nearly every case, one invention led to another and then to another. In fact, you can trace back the lineage of many modern items used today to the scientists and inventors explored in this book.

The Industrial Revolutionaries

The people you'll meet in this book are quite different from one another. Some were scientists who worked at universities; some were lone inventors who worked from their homes; some were even engineers who worked for large companies; and a few don't fit neatly into any category. But what these makers have in common is that they creatively used the technology available to them to make new, interesting, and valuable things.

Some of the makers you'll meet in this book are famous. You're almost certainly already familiar with Benjamin Franklin and his exploration of electricity, and you might also be familiar with Charles Goodyear and his process for making rubber. But some inventors are bit more obscure. Most people would not know of John Shore, who was an 18th century trumpet player with a scientific as well as a musical gift; or Louis Poinsot, a French mathematician who first laid out some of the important mathematical ideas we still use for designing buildings and ships today.

Besides Franklin, Goodyear, Shore, and Poinsot, you'll meet a couple of Italian scientists, Giovanni Venturi and Alessandro Volta, who stunned the European scientific community with pivotal inventions that led to things far greater than they could have ever imagined. And there's the steelmaker Henry Bessemer, who first figured out how to make steel affordably, and Joseph Louis Gay-Lussac who pioneered the art of making things fireproof.

Throughout the Industrial Revolution, great scientists all over the world figured out ways to make new things that made our lives easier, safer, and just plain better.

All of the makers in this book, who we'll call the *Industrial Revolutionaries*, made landmark discoveries during their careers that are still important in daily life. That's why it's worth knowing about them.

The projects I've included here are based on these inventions from the Industrial Revolution that propelled our understanding of science and technology forward.

There are many different ways to study these great inventions from the past. One way is to learn the history about how each was invented. This usually means learning about the inventors, where they lived, and how they went about their individual inventing processes; and, if you're really a scholar, finding out how they came up with ideas in the first place.

Another way to study an invention is to learn the scientific principles behind it. For instance, if the invention that you are interested in is the tuning fork, it would make sense for you to make sure you understand the science of acoustics, and how vibrations make sounds.

And a third way (and what I feel is the best way of all) is to re-create the invention yourself. This gives you an intimate understanding of how and why something works. So, if you take time to try a few of the projects I've included in this book, you'll get a special type of understanding that can only come to you by doing things yourself.

This book is about using all three ways—history, science, and building things ourselves—to become knowledgeable about many great inventions from the past and why they are still relevant today.

It was during the Industrial Revolution that the words *scientist* and *engineer* were first used, at least in the sense they are used today. If you were to go back in time and meet, say, Isaac Newton in 1687, or Benjamin Franklin in 1780, they would likely stare at you blankly if you called them scientists. Although the word sounds old and Greek, it's not. William Whewell, a Cambridge University professor in England, first coined the term in 1834. Before Whewell, the people we now call scientists were known as *natural philosophers*, or *cultivators of science*, or other rather unwieldy terms.

The title *engineer* also took on its modern meaning at this time—a person who is trained in designing mechanical or electrical technology. Before the Industrial Revolution, an engineer was generally thought of as a person who operated a siege engine or a catapult.

Based on the timeframe covered here, we'll meet and learn about quite a few people who were scientists and engineers before these words existed in their present form. In the chapters that follow, we'll take a brief look at the lives of these great inventors. Next, we'll examine their claim to fame; that is, we'll take a look at the wonderful or important thing they invented, the nature of the science behind that invention, and how it made the world a better place. Finally, and this is the best part, we'll build simplified versions of their inventions so we can really understand them and see for ourselves how and why they work.

It's worthwhile for you to also know that this book is part of a series of three. The first book, *ReMaking History, Volume 1: The Early Makers*, explores the contributions of great but ancient Makers who lived from the time of the earliest cave dwellers up to the Early Modern Era, which is just before the Industrial Revolution. This, the second volume, features the great inventors and creators of the Industrial Revolution, and the last book in the series, *ReMaking History, Volume 3: Makers of the Modern World*, covers the years from the turn of the 20th century until the present.

First Things First: Being Safe

I have designed the projects I describe in the following pages so that you can make and use them as safely as possible. However, as you try them out, there is still a possibility that something unexpected may occur. Many of these projects involve the use of nails, saws, glue, and power tools, and you need to be careful when you work with these items.

It is important that you understand that neither the author, nor the publisher, nor the bookseller can or will guarantee your safety. When you try the projects described here, you do so at your own risk.

The following are your general safety rules. You will find that some chapters also provide additional project-specific safety instructions.

1. Read the entire project description carefully before you begin the experiment. Make sure you understand what the experiment is about, and what it is that you are trying to accomplish. If something is unclear, reread the directions until you fully comprehend them.

2. Wear protective eyewear, gloves, and any other necessary protective clothing whenever indicated in the directions.

3. The instructions and information are provided here for your use without any guarantee of safety. Each project has been extensively tested in a variety of conditions. But variations, mistakes, and unforeseen circumstances can and do occur; therefore, all projects and experiments are performed at your own risk. If you don't agree with this, then put this book down—it is not for you.

4. Finally, believe me when I tell you that it's no fun getting hurt. I want you to stay in one piece. And the very best way to do that is to use your own common sense. If something doesn't seem right, stop and review what's happening. (That doesn't just pertain to what's in this book; that's my advice to you in general.) You must take responsibility for your personal safety and the safety of others around you.

Part I

Makers in Metal

On a clear, cold New England night in late 1807, those Yankees who were looking skyward observed a display of meteors unlike any they had ever seen. Not only were the heavenly streaks exceptionally bright, but they were noisy as well. Witnesses said they crossed the sky making a sound like "cannonballs being rolled on a wooden floor."

These were big meteors; in fact, some of them were large enough to enter the atmosphere and land without burning up. Meteorites fell upon large areas of Connecticut, Massachusetts, and Vermont. This was world-changing stuff, because prior to this, no one had ever considered that some of the rocks around us could have extraterrestrial origins.

The news of the rocks falling from above spread quickly, and two Yale University professors dashed out from New Haven to check things out. They spent days tracking down the landing spots and talking to locals. Several meteorites were brought to them for examination. After a great deal of careful inspection, they were ready to make a startling statement. These rocks, they declared, were not of this planet. They were, in fact, "stones from the sky."

At the time, this statement was so startling that almost no one believed them. Thomas Jefferson certainly didn't. It was "easier to believe that Yankee Professors would lie," he said, "than that stones would fall from heaven."

Nonetheless, these were indeed rocks from the sky, and given that such a large number of them were available for study, the scientists of the day had clear and unambiguous evidence that meteorites did exist. They were also able to draw conclusions about the composition of the fallen stars. Some of them, they found, were made of iron.

Now, this otherworldly iron was unlike anything seen on Earth. It was not simply a high-content ore, but rather, it was purified stuff—it was nearly 100-percent iron with a distinct crystalline structure.

Upon reading about this in scientific journals, archaeologists around the world sat bolt upright, for this finding could explain something that had been puzzling them for years.

For a very long time, scholars had been unable to explain the fact that evidence at archaeological sites had unequivocally showed the existence of very ancient, high-quality, iron tools. Up to now, this was a baffling find. At the time these tools were made and used, smelting or purifying iron was still an unknown skill. And on Earth, iron simply does not exist in a pure, unoxidized, metallic state.

These discoveries, found in several places on the planet, dated back as far as 4000 BC, which was thousands of years before the metal-refining breakthroughs of the Iron Age that began around 1300 BC. So how could such ancient people have tools made of pure iron?

But now that these meteors had been discovered, scientists had an answer. Could it be that those early iron artifacts were actually fabricated from extrater-restrial iron—meteoritic iron that quite literally fell from the sky? This theory now seemed quite logical, and years later anthropologists would find that ancient peoples had long used extraterrestrial iron blades on their bone tools and spears. In fact, an Inuit tribe in Green-land had chipped their iron for spear tips for generations, all from a closely spaced splash field of a huge meteorite now referred to as the Cape York meteor-ite (as shown below).

Like the tools of these indigenous peoples, the projects in this section uti-lize iron (as well as aluminum), although we'll procure our raw materials from less exotic sources than outer space. Metal-working is one of the foundations of the modern age, and the three projects that follow are both fun to make and fun to use; they include a rubber band gun, a tuning fork, and a campaign torch.

Henry Bessemer

Henry Bessemer and the Age of Steel

In this chapter, we'll discover how civilization moved from the Iron Age to the Age of Steel, and we'll explore the basics of heat treating and metalworking.

At the start of the Industrial Revolution, the state of the art in the field of iron metallurgy was the *puddling furnace*. Iron workers loaded crude iron ingots into the furnace and then continually stirred the molten metal with long iron rods through a small hole. This took great skill and practice; an experienced puddler was accepted in the industrial community as a highly skilled craftsman.

Puddling was an exceedingly difficult task. One scholar stated that "Only men of remarkable strength and endurance could stand up to the heat of the puddling furnace.... The puddlers were the aristocracy of the proletariat, proud, clannish, set apart by sweat and blood. Few of them lived past forty."

As the puddler stirred, solid chunks of iron would begin to appear in the liquid mass. He'd gather these chunks and work them under a forge hammer, yielding a slab of hot wrought iron. That slab could then be run through rollers in rolling mills to form flat iron sheets or rails.

Although the process was hard, cumbersome, and expensive, the end product was wrought iron, not steel, so it did not have the strength or utility of steel. *Steel* is iron alloyed with carbon, and it is superior to plain iron in every way. But prior to 1856, there was no practical way to control the percentage of carbon in iron, so there was no way to manufacture steel at a price the industry of the time could afford.

The Dawning of the Age of Steel

One of the things that drove the rise of the Age of Steel was the railroads. They were booming in the middle of the 19th century and, because of that, so was the market for iron and steel. The first railroads ran on wrought iron rails that were too soft to be durable. On some busy stretches, and on the outer edges of curves, the wrought iron rails had to be replaced every six to eight weeks! Although the railroads knew that steel rails would be far more durable, there were no practical manufacturing methods to produce them affordably.

It was at this point that a smart fellow named Henry Bessemer came on the scene. Bessemer, an engineer and metallurgist, would, over the course of his career, receive 129 patents in a variety of engineering disciplines. But the invention for which he was knighted and the one that made him a very rich man was the one that involved turning iron and coke into steel.

Henry Bessemer and Carbon Steel

While looking for a way to increase the strength of iron cannon barrels, Bessemer discovered that carbon, if dissolved within molten pig iron, unites readily with oxygen. Knowing that, he determined that if he could blast a jet of air through molten iron, then he could convert the pig iron into much stronger alloy steel by accurately controlling its carbon content.

Bessemer decided to test his theory. He built an experimental furnace in the backyard of his laboratory in London. He constructed a high-temperature heating chamber that was 4 feet high and had a 12-horsepower steam engine to run the air injector. When the pig iron in the chamber liquefied and he turned on the blower, a fireball erupted from atop the molten iron. But when he sluiced the molten metal into ingot molds, he gazed with delight upon the "limpid stream of incandescent malleable iron almost too brilliant for the eye to rest upon." It was then that Bessemer knew he had found a way to make cheap steel.

In 1856, Bessemer designed what he called a *converter*—a large, napiform receptacle with holes at the bottom where pumps could inject compressed air. Bessemer filled the converter with molten pig iron, blew compressed air through the molten metal, and found that the pig iron was indeed emptied of excess carbon and silicon in just a few minutes. From that time onward, carbon steel could be made abundantly and cheaply—the Age of Steel had begun.

Materials

Music wire, ³/₃₂" diameter × 36" length (most hardware stores sell this tough, high-carbon steel wire, sometimes called piano wire)

Heat-shrink tubing, 8" long

Rubber bands, assorted sizes and widths

(1) Quart of soybean oil

Tools

Propane torch

¾" steel pipe about 1' long (¾" steel pipe has an outside diameter of about 1")

Large needle-nose pliers

Slip-joint pliers

Safety glasses

Heavy gloves

Hot plate

Candy thermometer

Medium pot with lid

Hacksaw or rotary cutoff tool

Metal vise

Making a Steel Rubber Band Gun

Steel is such an important raw material because it is strong, yet malleable. Further, steel can be heat treated in many ways to make it harder or softer, flexible or stiff, or ductile or brittle, depending on the application for which it is needed. In this exercise, you will heat treat steel to make a rubber band gun. To do so, you first need to anneal a piece of steel music wire to make it soft and malleable, then you need to cut and form it to shape, and finally, you need to temper it to give it the tough, springy characteristics your gun requires.

Building the Rubber Band Gun

Owning a self-made rubber band gun puts you near the top of the office or school-prankster hierarchy. But remember, with great power comes great responsibility.

1. For this activity, the first thing you need to do is don safety glasses and heavy gloves. See "Hot Stuff! Keep Safe." at the end of this procedure for the reasons why.

2. Once you're properly protected, place one end of the music wire in the vise. At this point, you'll notice that the wire is very stiff and difficult to shape.

3. Light the propane torch and heat the wire until it is cherry red (see Figures 1.1 and 1.2).

Figure 1.1: Heating the wire with a propane torch

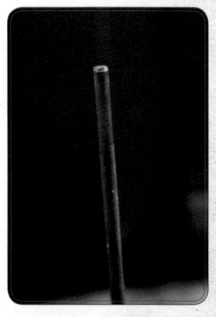

Figure 1.2: The cherry red wire

Cherry red is a fairly dull color. If your wire starts glowing bright orange, you've heated the metal too much.

The heating to cherry red and subsequent air cooling of the wire is a process called *annealing*, which makes the stiff music wire soft and malleable.

4. Place the steel pipe in the vise and form the cooled wire around the pipe to make the round top and bottom springs of the gun, as shown in Figure 1.3.

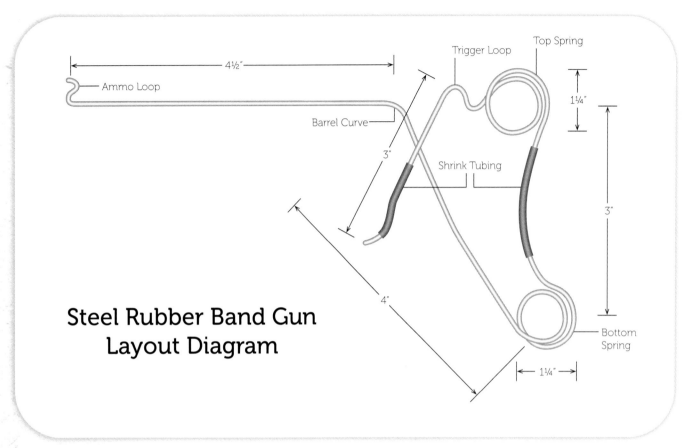

Figure 1.3: Making the springs of the rubber band gun

5. Next, use the pliers and vise to bend the ammo loop, the barrel curve, and the trigger loop into the shapes shown in Figure 1.3.

6. When you finished these bends, cut off any excess wire with a hacksaw or a rotary cutoff tool.

7. Heat the top and bottom springs of your gun with the torch until they glow cherry red. Once they're hot, plunge them into a bowl of water.

8. Dry the springs and clean off the scale with steel wool. Be careful—at this point, the steel in these springs is very hard and extremely brittle. Bending them even a little will cause them to break.

9. Pour the soybean oil into the pot, place it on the hotplate, and heat it to 400°F. Use the candy thermometer to check the temperature.

   Again, use caution here—the hot oil may smoke, so do this step outdoors or keep a window open and ventilate the room.

10. Reheat the round springs with the torch until they glow cherry red. Stop heating them and allow them to cool until the red color is completely gone; then plunge the springs into the oil.

11. Turn off the hot plate and allow the oil and the springs to cool slowly to room temperature. This slow annealing process will restore the temper to the steel (see Figure 1.4).

12. After the oil and newly annealed steal are completely cool, wipe off the oil and thread a piece of shrink tubing over

Figure 1.4: Quenching the springs in the oil

the handle and trigger of the gun; then shrink the tube using a match or heat from the hot plate.

If the heat treating process worked properly, you now have a springy steel rubber band shooter (see Figure 1.5). Load up and pull the trigger!

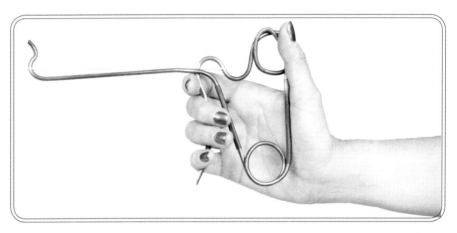

Figure 1.5: The completed rubber band gun

Hot Stuff! Keep Safe.

When you are working on this activity, please make sure you don't get injured by following these safety tips:

1. Use extreme care when handling the hot oil and wire.

2. The smoke point—that is, the lowest temperature at which smoke begins to rise from the surface of the soybean oil—is quite high, around 450°F. And even then, the oil doesn't ignite readily. Still, take care to

   - Not let the oil heat up past the smoke point.

   - Have a lid handy to cover your pot in the unlikely event of a fire.

   - Keep a grease-fire-capable fire extinguisher (check the fire extinguisher label to make sure) close by.

How Bessemer Converters Work

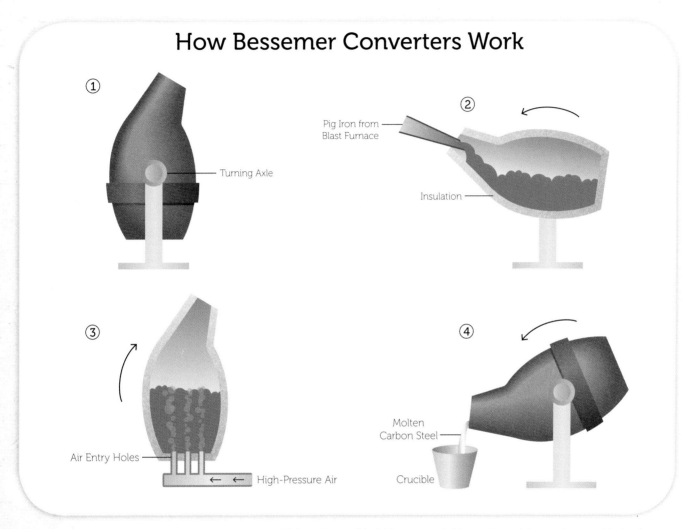

The Bessemer Converter is a rotating insulated container (1) that takes molten high-carbon pig iron and turns it into carbon steel. In the first part of the process, the converter is tilted so molten iron can be loaded, as shown in drawing (2). Then high-pressure air (3) is forced into the bottom of the upright converter. The air removes impurities and excess carbon, turning iron into steel. Finally (4), the converter is tilted and the molten steel flows into crucibles.

Figure 1.6: Turning pig iron into carbon steel

John Shore Invents the Tuning Fork

2

The science of acoustics is an ancient one, and historians of science trace its origin to the ancient Greek island of Samos, the pinnacle of classical antiquity's Ionian culture. It's there that the scientific principles for acoustics and sound were first discovered. That's not to say it's where the first musical instruments were built, for music goes back into Paleolithic times, but it is where the initial understandings of how and why sounds, both pleasant and unpleasant, are made.

Pythagoras of Samos, the famous Greek philosopher of the mid 4th century BCE, began to experiment with a vibrating string, and as far as we know, he was the first person to rigorously study the relationships between mathematics, science, and music.

The most popular instrument in Pythagoras' Samos home was the lyre. This early harp consisted of a wooden frame on which were strung four strings. Each string was tensioned to produce a certain pitch when plucked. Being a music lover as well as a thoughtful fellow, Pythagoras was intrigued by this instrument; by studying it, he came to understand how the lengths of strings produced pitches, and further, he recognized some string lengths sounded better than others, at least in relationship to the pitches that preceded and followed them.

The discoveries of Pythagoras launched a thousand musical explorations and, over time, the science of acoustics has become well understood.

Inventing the Tuning Fork

After much of the city was destroyed by fire in 1666, London was elegantly rebuilt, and by the end of the 17th century, it was perhaps the most prosperous city in Europe. Occurring simultaneously with this rebirth was a cultural revival that placed great emphasis on music. Music was being played constantly in every corner of the city, and it was at this time that many of England's most well-known composers—George Frideric Handel, Henry Purcell, and Jeremiah Clarke, among others—worked and performed.

One talented young trumpet player, John Shore, was feted by royalty and commoner alike. As his star ascended, he was named Sergeant Trumpeter to the English Crown, a lucrative position that allowed him to collect and pocket license fees from nearly every trumpet player in England.

By 1711, the multitalented Shore was a rich man who was able to afford a bit of leisure time. Having a bit of the inventor in him, he decided to investigate alternatives to the pitch pipes that musicians of the time relied on to adjust their instruments so they could play in tune with other musicians. The pitch pipes of that era were often unreliable because they were much affected by changes in temperature and humidity. Because of England's damp and changeable climate, musicians who relied on pitch pipes often found themselves playing out of tune.

Shore devised a new method for tuning instruments that he called a *tuning fork*. This metal fork is impervious to weather and provides the same clear, constant tone in all conditions. When Shore was asked to help tune an orchestra assembled for a concert, he would often say, "I have not about me a pitch-pipe, but I have what will do as well—a pitch-fork." He is said to have laughed heartily at his pun.

Making a Tuning Fork

The wonderful thing about a tuning fork is that it is fairly simple to make, although you must take great care in dimensioning it. In fact, if you have a small strip of aluminum or brass, you can fabricate a tuning fork to strike a particular tone (say, the one called Concert A, which vibrates at 440 hertz and is the note orchestras tune to) with just a few tools and a little elbow grease.

The pitch of a tuning fork is a function of its material composition and the dimensions of its *tines*—the metal protrusions attached to the handle. In fact, it is possible to calculate the correct size and shape of the tines to attain particular pitches. But unfortunately, I won't be discussing these equations in this chapter because they get complicated in a hurry; not only do they contain some higher-order mathematical terms, but they also require you to know several material properties, including the modulus of elasticity and density and the moment of inertia for the shape of the tines.

But—good news!—all is not lost for the aspiring tuning-fork maker; you can fabricate a reasonably accurate fork using trial and error. The rule to remember is that longer tines vibrate more slowly and produce lower-pitched tones. So, if you start with tines that are slightly longer than you expect you will ultimately need, you can simply file off material until the pitch reaches your desired value.

The following construction directions talk you through making a fork that produces a pitch close to Concert A. To fine-tune it, simply listen to the pitch it makes and compare it with a tuned piano or a smartphone tuning app; then incrementally file away aluminum from the ends of the tines until the pitch is just right.

Materials

(1) 9"-long strip of aluminum, ⅛" thick by 1½" wide

Tools

Hacksaw

Electric drill

Workbench with vise

Flat file

Pencil

Ruler

How to Build Your Tuning Fork

Fabricating your own tuning fork is fun and easy, although it takes patience and care to make it sound accurate and clear.

1. Use a pencil to lay out cutting lines on your strip of aluminum. Tines that are ½-inch wide and about 6⅛ inches long will produce a sustained tone of 440 hertz (the middle A on a piano keyboard). Mark what will become the waste pieces with Xs or other lines so you are certain what you want to keep and what you want to save for some other purpose. Figure 2.1 gives you an idea of what you're shooting for.

Figure 2.1: Diagram of tuning fork

2. Place your aluminum strip securely in the vise of your workbench and cut the aluminum strip to shape using the hacksaw (see Figure 2.2). This can take awhile depending on the keenness of your saw blade and the vigor with which you wield the saw.

3. Use an electric drill to cut small holes in the end of the center waste piece so you can more easily remove it (see Figure 2.3).

Figure 2.2: Cutting the aluminum strip with a hacksaw

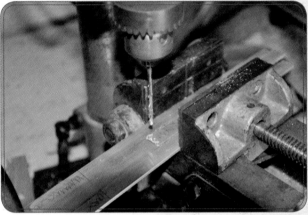

Figure 2.3: Drilling holes to make scrap piece easier to remove

4. File the fork so the edges are smooth and even (see Figure 2.4).

Note: In order to obtain a clear, single tone, each tine must be exactly the same dimension. In order to obtain a desired frequency, both the thickness and length of each tine must be exact.

Figure 2.4: Filing the tines

Materials

**(4) Wood boards,
10"×5½"×½"**

**(1) Wood board,
5⁷/₁₆"×4⁷/₁₆"×½"**

Wood glue

Tools

Electric drill

Building a Resonator Box

In order to hear the tuning fork more clearly, it helps if you build a resonator box.

To build the box, follow these steps:

1. Join the four pieces of wood with glue to form a square box with open sides, as shown in Figure 2.5.

2. Let the glue dry.

Figure 2.5: A resonator box and tuning fork

3. When all the glue is dry, drill a hole in the top of the box that will hold your tuning fork firmly in place.

4. Insert the handle of the fork into the hole.

   Now when you tap the fork, you should hear the tone resonate.

A resonator box works most efficiently when the length of the open box is one quarter the length of the tuning fork's natural wavelength. You can look up the frequency and wavelength of any musical note in an acoustical handbook or simply by searching for this information on the Internet. For example, given that the wavelength of a 440-hertz (Concert A) tone is 30.9 inches, a resonating box for a Concert A tuning fork should be $30.9 \div 4 = 7.7$ inches deep in order to get the strongest tone.

When you attach your tuning fork to a wooden box with the same resonant frequency, the energy from the vibrating fork enters the box and causes it to vibrate at the shared natural frequency. The resonator box, in turn, causes nearby air molecules to vibrate at the same frequency as the fork and box. Physicists call this *forced vibrations*. Because the wooden box has more surface area exposed to the air molecules than the fork, when you attach a resonator box to the fork and strike it so it vibrates, this causes the amplitude of the sound to increase.

Tuning Fork Science

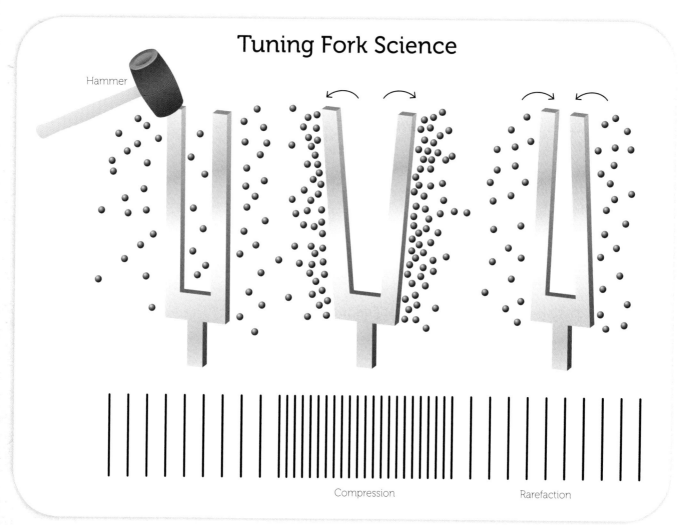

Compression

Rarefaction

When you tap your tuning fork on the table to make it ring, the tines begin to vibrate—that is, they move in and out from their resting position. When they move out, the tines push against the air molecules, compressing them into a smaller volume, and that forms, a temporary region of high pressure. And then the tines move back in, causing a void of molecules in that same space, and that makes a low-pressure region. This happens hundreds of times per second (440 times per second, for a tuning fork tuned to concert A).

Figure 2.6: The science behind a tuning fork

How a Tuning Fork Works

The sensation we term *sound* is caused by movement, specifically, the movement of something in air. If you think about it, you'll always find a repetitive motion or oscillation at the origin of every audible noise. For example, the sound that a piano makes comes from its vibrating strings. Even where the vibration cannot be seen, as in say, a clarinet, it may still be felt. Figure 2.6 explains how the concept works for the tuning fork.

Abraham Lincoln

Abraham Lincoln and the Campaign Torch

3

Although it is a simple device, the torch has a uniquely long-lived importance in human history. Few things that we still use today were invented so long ago. In fact, in some parts of the world, the fire from a flaming torch was the dominant source of illumination for 99 percent of human history.

Even today, from the Olympic Games to the Statue of Liberty, the torch remains one of the most important and significant symbols in cultures throughout the world. Invariably, it remains a positive symbol, representing life, hope, and goodwill. Maybe that's why torches were once so important to the American political process.

The Light of Politics

Presidential political campaigns were much different in the 19th century than they are now, and to many people (including me), they sound like much more fun. Instead of ceaseless televised debates and commercials, scripted sound bites, and never-ending media analysis, the key political tool was the parade.

Although everyone may still love a parade, Americans of 150 years ago, it seems, were absolutely enamored of them. Imagine for a moment you are member of the Wide Awakes, one of many political marching clubs organized to drum up support for political candidates. It's a pleasant autumn night in 1860, and word has been received that a march on behalf of your presidential candidate, Abraham Lincoln, has been organized. For the politically active person, this is terrific news!

Since marching is what you like to do, you and your fellow Wide Awakes do it often and are very good at it. Everyone in the group (and there are thousands) owns a torch. Campaign torches, thanks to the manufacturing techniques of the Industrial Revolution, are quite cheap, but if you want something special, you have to spend a quite a bit more. Your torch—a new gimbal-mounted, nickel-plated, tin torch in the shape of a musket—is particularly eye-catching.

In the evening, the Wide Awakes, as do all political marching clubs, wave their torches with pride and artistry; they even use them in the manner of rifles, presenting a display of close-order drill to the crowds lining the streets. It's very exciting:

> *Thousands of torches flashing in high, narrow streets, crowded with eager people and upon house-fronts in which every window swarms with human faces, the rippling, running, sweeping and surging sounds of huzzas from tens of thousands, with the waving of banners and moving transparences of endless device are an imposing spectacle and these everyone in the city saw at the Wide Awakes festival on Wednesday night.*
>
> —Harper's Weekly, October 13, 1860

Parades often lasted two to three hours. The costumed or uniformed participants sang campaign songs and shouted slogans as they marched. To satisfy the need for the thousands of torches that accompanied such parades, scores of small manufacturing companies sprang up across the United States to fabricate parade torches. They made torches in many shapes and forms, ranging from rifle look-alikes for the afore-mentioned close-order drill ceremonies to torches built in the shape of faces, animals, capital letters ("L" for Lincoln," for instance), hats, pinecones, brooms, and pick axes.

Mr. Lincoln himself rarely attended actual parades, because at the time, candidates did not campaign personally. They stayed home and let others make speeches on their behalf. But on August 8, 1860, Lincoln did participate at a rally near his home in Springfield, Illinois (see Figure 3.1). He was mobbed by an enthusiastic crowd and was lucky not to have been injured.

Torchlight parades as a political campaign tool were extremely popular in the second half of the 19th century and, accordingly, it was a boom time for torch manufacturers. Their factories ran at full steam, stamping out hundreds of thousands of unusually shaped torches for every closely contested election. Night after night, all over the country, people marched by torchlight, hoping the bright lights held aloft would awaken sympathetic feelings in onlookers and carry their candidate to victory.

But the era of such campaigning tactics was soon to wane. In the 1860s and 1870s, strategies such as parades were the best way to reach people of all social status, literate or not. But as literacy rates rose and newspapers became less politically biased, at least overtly, political campaigning became less spectacular and more educational. By 1900, the importance and frequency of the torchlight parade declined dramatically, and the torch manufacturing industry slid into a steep decline from which it never recovered.

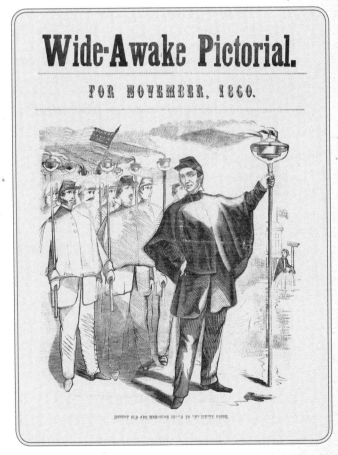

Figure 3.1: Lincoln carrying his own torch

Materials

Drill with ⅝″ and ³⁄₁₆″ bits

(1) Empty and clean small metal can with screw-on or replaceable push-on lid

(1) #10 round-headed wood screw, ¾″ long

High-temperature epoxy adhesive

(1) ⅝″ hex nut

(1) Piece of ½″-diameter cotton rope, 2½″ to 4″ long depending on can height

(1) 1″-diameter wooden dowel, 3′ long

Kerosene (Do *not* use gasoline or alcohol!)

Fill spout for the kerosene

Aluminum foil or high-temperature aluminum tape

Long-handled lighter or fireplace match

Fire extinguisher

Making a Parade Torch

No 19th century political candidate worth his salt would allow his supporters to march on his behalf without making sure a blazing torch was in everyone's hands.

Here's what you'll need to make your own.

Building the Lincoln Torch

Refer to Figure 3.2 for an overall understanding of the project.

1. Drill a ⅝-inch hole in the lid of the clean metal can as shown in Figures 3.3 and 3.4.

How to Make a Parade Torch

Drill a ⅝″ hole.

Epoxy a ⅝″ hex lock nut over the hole.

Insert a ⅝″ rope that is 2½″ long.

Open the lid to fill the reservoir with kerosene.

Epoxy the can to the pole. And use a #10 wood screw.

Form an aluminum foil skirt.

Light the torch carefully.

Figure 3.2: The assembly diagram

2. Using high-temperature epoxy, make a wick collar by gluing the hex nut over the hole as shown on the bottom left of Figure 3.2 and in Figures 3.5 and 3.6.

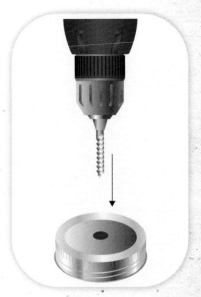

Figure 3.4: Drilling a ⅝-inch hole in can lid

Figure 3.3: The drilled hole

Figure 3.5: Gluing the hex nut to the lid

Figure 3.6: Using the epoxy, glue the nut to the lid.

3. Drill a ³⁄₁₆-inch-diameter hole in the bottom center of the metal can.

4. Fasten the can to the top of the wooden dowel with the #10 screw as shown in the center of Figure 3.2 and in Figures 3.7 and 3.8.

5. Using high-temperature epoxy, seal around the screw so the can is well attached to the wooden dowel and leak proof.

6. Let the epoxy harden before continuing. Check the epoxy's label directions for curing time.

Figure 3.7: Attaching the can to the dowel

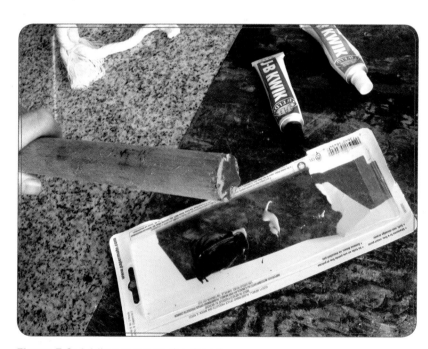

Figure 3.8: Adding epoxy to the dowel

7. Cut your rope to the proper length based on the height of your can (see Figure 3.9) and then insert the rope through the hex nut in the can lid (as shown in the center of Figure 3.2). It should fit snugly with about a ½ inch of wick sticking up.

Figure 3.9: Cutting the rope

8. Cover the can with the aluminum foil or high-temperature aluminum tape, forming a skirt around the can (see the right side of Figure 3.2).

Keeping Safety in Mind

As you're working through this project, make sure to keep the following in mind:

- Use your torch only outdoors and only where it can't accidentally start a fire.

- Use only kerosene. Although kerosene is not as flammable as gasoline, you still need to be extremely cautious. Make sure you store it in an approved container.

- Keep a fire extinguisher handy. Be careful when you are lighting, handling, filling, or holding the torch. Never fill the torch while it is hot.

- Check often to make sure the can is securely attached to the dowel.

Using the Torch

When you are getting ready to use the torch, follow these steps:

1. Fill the torch a third of the way full with kerosene (see the center of Figure 3.2). Make sure to do this outdoors using a fill spout.

2. Make absolutely sure that the lid is pressed down securely or screwed tightly after you fill the torch.

3. Let the rope wick draw kerosene up. After one to two minutes, light the wick using a long-handled lighter or fireplace match. Your torch should light easily (see Figure 3.10).

Figure 3.10: The lit torch

4. When you're lighting the torch, do not tip the torch too much or it might drip kerosene.

5. If you'd like, you can whittle the other end of the dowel to a point so you can place the torch in the ground in your backyard (see Figures 3.11 and 3.12).

Figure 3.11: The backyard torch

Figure 3.12: The backyard torch at night

The Science Inside: In the Tip of a Flame

When the bottom of a rope wick is immersed in a reservoir of kerosene, the fuel molecules begin to travel up the wick to its tip via a plethora of chemical processes, including capillary action, cohesion, and adhesion. (If you've read *ReMaking History, Volume 1: The Early Makers*, then you've already read about how these processes work in the chapter about oil lamps.) Before long, the wick is wet with kerosene from top to bottom and is ready to be lit.

When you bring the high heat of a lit match head into contact with the rope wick, enough additional energy is added to shake loose some of the kerosene molecules at the tip of the wick. These gas molecules change to vapor as the gaseous kerosene continues to heat in the match flame until so much energy is absorbed that the kerosene's molecular structure breaks down. As it does, the chemical bonds holding the individual atoms in each molecule are ripped asunder, which causes the vapor to change from a cloud of long, complex hydrocarbon molecules into a mist of simpler, but far more reactive, fragments of carbon and hydrogen.

This process, called *pyrolysis* or "cracking," is the hidden mid-step between fuel and flame, the secret transformation that kerosene, coal, candle wax, or indeed, any burnable fuel goes through before it chemically breaks down to create the heat and light of fire.

The hot mist of reactive fragments spreads outward, soon colliding with oxygen in the air just beyond the immediate area of the wick. Now combustion takes place in earnest; carbon and hydrogen atoms quickly recombine to more and more stable forms until all that remains are (chemically speaking) the rock-solid molecules of water and carbon dioxide (CO_2).

Basically, the oxidation process starts with long, energy-rich, but unstable, molecules of kerosene, and it culminates in small, energy-poor water and CO_2 molecules. Along the way, the fire reaction produces heat and light. Figure 3.13 shows this process and explains how by-products of the reaction form.

Science Explainer: Why Soot Forms

Scientists call the flame that your torch produces a "laminar diffusion flame." This is the same sort of flame you'll find in a wood-burning fireplace or in a wax candle. In the wick of your torch, there are a number of processes happening concurrently, and some of them are very complicated. But simply put, fuel travels up the wick by capillary action. At the tip, oxygen combines with the fuel through a random-mixing process known as diffusion.

Diffusion flames are slow-burning flames, and they often produce more unburned fuel leftovers, called soot, than the faster burning flames in propane torches or butane lighters. That's because the mixing process is more or less random, so some fuel molecules do not find sufficient oxygen for the reaction to go to completion.

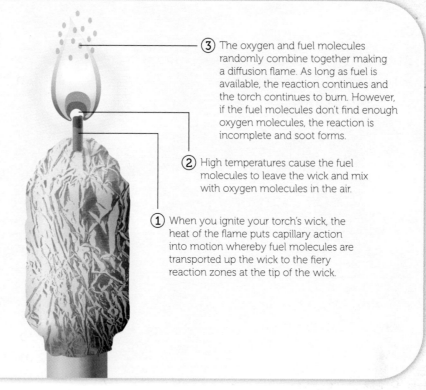

③ The oxygen and fuel molecules randomly combine together making a diffusion flame. As long as fuel is available, the reaction continues and the torch continues to burn. However, if the fuel molecules don't find enough oxygen molecules, the reaction is incomplete and soot forms.

② High temperatures cause the fuel molecules to leave the wick and mix with oxygen molecules in the air.

① When you ignite your torch's wick, the heat of the flame puts capillary action into motion whereby fuel molecules are transported up the wick to the fiery reaction zones at the tip of the wick.

Figure 3.13: From fuel to fire

Part II

Physics and Chemistry Revolutionaries

The Greek philosophers of the Classical era, led by Empedocles, who was followed by Aristotle, believed that everything in the universe was made up of four elements: earth, air, water, and fire.

Everything in the world, from corn to cotton to clouds, was composed of these four elements; the differences between all the things we see or touch was only the relative abundance and mixing manner of the four constituent components. For instance, clouds, according to Aristotle, were a light composition of air and water. Cotton, on the other hand, was a mixture of rock, air, and water, and differed from corn only by the difference in proportion of each element. This thought pattern was known as the Four Element hypothesis, and it was accepted with only minor modifications as the cosmological basis for the entire universe for more than a thousand years.

But, starting in the late Middle Ages, this explanation no longer fit the results of many experiments. Although true scientists, as we know them now, did not exist, there were alchemists aplenty, and many of these experimenters actually had excellent analytical technique. In their quest to find a way to turn common metals into gold, the alchemists analyzed a number of chemical processes; these experiments produced results that didn't jibe with the classical Four Element worldview. At this point, the classical Greek pillars of cosmological doctrine began to crumble. The alchemists were on to something, and they were beginning to suspect that the world was much, much more complex than they had been led to believe.

By the 1700s, new ways of thinking—scientific instead of alchemical—took hold. Real science came into existence, starting with physics, followed by chemistry. The great early thinkers—Isaac Newton, Robert Boyle, John Dalton, and Antoine Lavoisier among them—were able to piece together the fundamental truths of the way the world really worked.

Giovanni Venturi

Giovanni Venturi and the Venturi Effect

4

Giovanni Battista Venturi was an 18th century Italian polymath who did a remarkable number of things extraordinarily well. He was an ordained Catholic priest, a university-level mathematics instructor, one of the leading civil engineers of his day, a politician and statesman (and a favorite of Napoleon Bonaparte), and a world-class historian. In fact, it was Venturi who first called attention to Leonardo da Vinci's scientific contributions. Although we honor da Vinci today for being both a brilliant inventor and artist, prior to Venturi's scholarship, da Vinci's contributions to science were largely unheralded.

Perhaps Venturi's greatest impact, though, came from his trail-blazing research in the field of fluid mechanics. The Venturi effect—the eponymous scientific concept that he first wrote about in the late 18th century—is the principle on which things as varied and important as paint sprayers, fertilizer applicators, gas grills, and scuba regulators are based.

In his 1797 book, Venturi describes how the motion of one fluid can "impress its motion on other fluids, by carrying them along in what I call the lateral communication of motion in fluids." Venturi discovered that air or water shooting through a constriction in a pipe can more or less magically drag another pipe full of fluid along with it if the geometry of the pipes is just right.

The Venturi Effect

Venturi really didn't understand the reasons behind this phenomenon, now universally known as the *Venturi effect*. But it is easily explainable today using the important law of fluid mechanics known as the Bernoulli equation. Although the Bernoulli equation is a bit complex and requires more space to explain than can be given to it here, the Venturi effect is relatively easy to understand without this math. Stated in broad terms, fluids that are under pressure and are moving through a gradually narrowing pipe gain speed. That's easy enough to understand; since the number of fluid molecules going in and coming out of the pipe is constant, the fluid molecules have to speed up in order to move through the constriction.

The insightful thing that Venturi noticed is that as the fluid flowing through the pipe constriction speeds up, concurrently, the pressure the fluid exerts on the pipe walls drops. In fact, if you measure the pressure the fluid exerts at various points in a system of converging and diverging pipes, you find that the pressure is the smallest where the speed and pipe constriction is the greatest, as seen in nature in the stream diagram in Figure 4.1.

When you first consider the information in Figure 4.1, this bit of knowledge may seem a bit ho-hum, but it's actually a huge insight—a scientific and technological bonanza, in fact. It turns out that using this effect is actually a really neat way to move, push, drag, or mix one fluid by using nothing but the motion of another— this is Venturi's "lateral communication of fluids."

The Venturi Effect is a special case of a physics law called Bernoulli's principle. When you use your thumb to partially block the opening on your garden hose, the restriction makes the water come out faster. The scientific explanation for why this happens is Bernoulli's principle. Basically, this idea, first teased out by Swiss physicist Daniel Bernoulli, states that as the speed of a moving liquid or gas increases, the pressure within the fluid decreases.

You can see the same thing happening to the speed of the current in a river when the width of a river changes. Water running through the wider river section travels slowly, but it speeds up through the narrower parts. Bernoulli's principle says that the pressure within the fluid decreases as the water speeds up. Likewise, the pressure in the fluid increases as the water slows down in the wider regions. This idea may seem hard to understand. In fact, you may be thinking that since the water in a pipe or river restriction is in a tighter space, the pressure should increase. Well, it does, but it's not the pressure within the fluid that increases. The pressure increase is felt by whatever surrounds the fluid. In a pipe, the walls of pipe feel more pressure in the narrower part. In the river example, the pressure on the riverbanks increases in the narrow sections.

Figure 4.1: Velocity versus pressure in a moving stream

Of its countless applications, the Venturi effect is probably best known for its role in the automobile carburetor. In a carburetor, air flows through a Venturi channel; simultaneously, gasoline is sucked through an opening in just the right proportion to the air. The gasoline, drawn in by the Venturi suction and combined with air, goes into the car engine's cylinders; there, the spark plug ignites the mixture, which pushes down the engine's pistons and ultimately makes the car move forward.

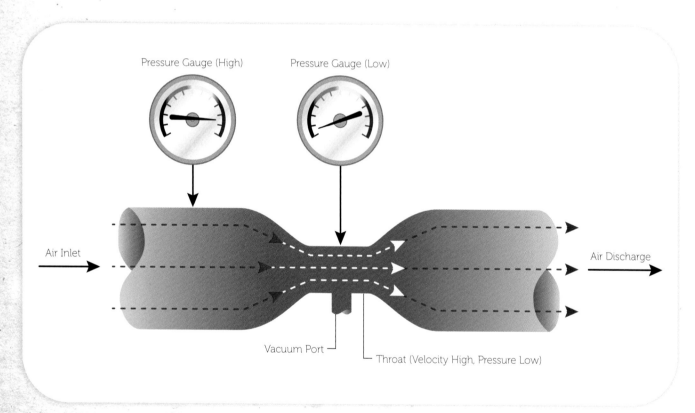

Figure 4.2: Diagram of a Venturi vacuum generator

Making a Venturi-Style Vacuum Pump

Venturi's lateral communication of fluids can also be used to make a simple, but effective, vacuum pump. Although there are many different types of vacuum pumps (see Chapter 10 of *ReMaking History, Volume 1: The Early Makers*), the Venturi effect–powered vacuum pump is certainly the simplest. Imagine a pump with no moving mechanical parts that is extraordinarily simple and cheap to make!

Here, we will make a Venturi-based vacuum pump like the one in Figure 4.2 out of materials that you can find for less than $25. Vacuum pumps are pretty awesome science tools and they lend themselves to quite a variety of interesting experiments.

Although I don't think this project is particularly dangerous, make sure you are aware of the following two pieces of information. First, you'll need access to an air compressor in order to speed the air through the Venturi system, and you must always handle high-pressure air with caution. So, make sure you use your air compressor in accordance with the manufacturer's directions and make sure you wear protective eyewear. Second, when you draw a vacuum in a container, be aware that the pressure the atmosphere exerts could cause the container to implode. For items like plastic soda bottles, that's pretty cool. But for other items, like glass bottles, that's not so good. Before you apply your pump to a glass flask or bottle, be sure it is strong enough to handle an atmospheric vacuum.

These are the supplies you will need for this project.

Materials

Vacuum-worthy plastic or glass bottle or flask with air-tight lid or stopper

Air compressor with hose terminating with a ¼" female industrial-style coupler

Thread compound

¼"-diameter soft, flexible plastic tubing, about 18" long

¼"-diameter plastic or brass double hose barb

Venturi vacuum generator—mass-produced ones are very inexpensive, are sold by many online vendors, and work quite well. I bought one on sale from Harbor Freight Tools for $19, and that included the air hose and tube fittings I needed. Search on the Internet for "Venturi vacuum generator" or "Venturi vacuum pump."

Tools

Air compressor

Adjustable wrench

Blade-style screwdriver

Constructing the Vacuum Pump

A vacuum pump is the opposite of a normal air pump. It removes air from a closed container instead of putting it inside. There are a number of scales used for measuring vacuum such as "inches of mercury," torr, and pounds per square inch (psi), but they all indicate how much air has been taken away from a particular space.

1. The first thing you'll need to do is remove the vacuum module from the vacuum pump plastic housing (see Figure 4.3). To do so, use your adjustable wrench to unscrew and remove the male air connector.

2. Then, remove the tee fitting from the brass nipple.

3. Now use a screwdriver to separate the two halves of the plastic housing; remove the plastic housing pieces and discard.

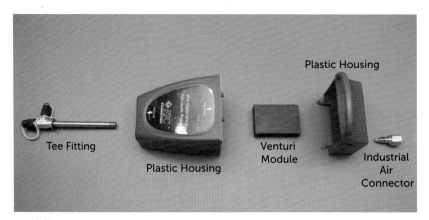

Figure 4.3: The disassembled vacuum pump

4. Now reattach the tee fitting and the industrial air connector onto the Venturi module (see Figure 4.4).

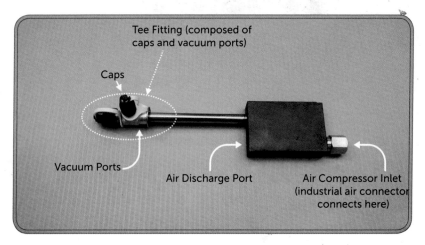

Figure 4.4: The reassembled vacuum pump

5. Use thread compound on the threads and torque the fittings down using the adjustable wrench to make sure there are no leaks.

6. Carefully work one end of the ¼-inch soft plastic tubing over the smaller of the screw thread fittings that is attached to the pipe nipple.

7. Securely cap the larger fitting with the provided cap (see Figure 4.5).

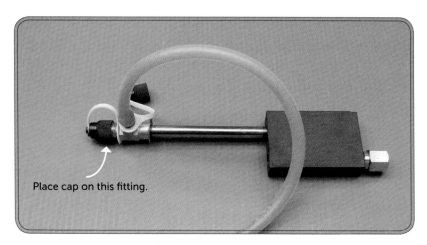

Place cap on this fitting.

Figure 4.5: Hose and cap added

8. Insert one end of the double hose barb into the remaining open end of the soft plastic tubing (see Figure 4.6).

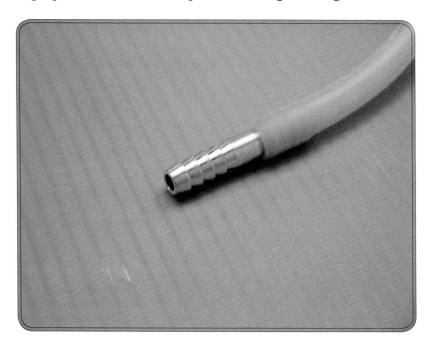

Figure 4.6: The inserted barb

9. Now attach the inlet of your vacuum pump to an air compressor (see Figure 4.7).

Figure 4.7: Attaching the pump to the air compressor

10. To operate your vacuum pump, set the discharge pressure of your air compressor to between 70 and 90 psi. The higher the pressure, the greater the vacuum your vacuum pump will draw. At 90 psi, I obtained a vacuum around 24 inches of mercury. That's a lot.

Fun with a Vacuum Pump

Now that you've assembled a vacuum pump, what will you do with it? You can attempt many interesting science experiments. The easiest is to drill a hole in the plastic top of a soda bottle that is slightly larger than the tip of the hose barb and insert the barb in this hole while the air compressor (and thus the pump) is running. It's oddly satisfying to implode bottles and then watch them regain their shape when you remove the vacuum hose.

Another very easy trick is to place a few marshmallows or a dollop of shaving cream inside a vacuum-capable flask (see Figure 4.8). Place a single-hole stopper on the flask, pull a vacuum, and watch what happens to the stuff inside.

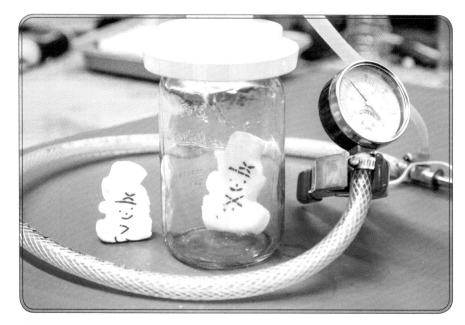

Figure 4.8: Experimenting with vacuums and marshmallows

Another classic science experiment is to place a piezobuzzer inside a container and then evacuate the air. When all (or as much as is possible) of the air is removed, the buzzer will go silent or nearly silent. This is because the vacuum can't propagate sound waves. In Chapter 2, "John Shore Invents the Tuning Fork," we discussed the fact that sounds are basically vibrations that are propagated through air. If the air is removed from a container there is no vibration and thus, no sound.

If you're really ambitious, you can take a crack at sous-vide cooking. (*Sous vide* is a French culinary technique that involves the artful preparation of foods while they are under vacuum. It involves vacuum-sealing ingredients in plastic and then cooking them in a temperature-controlled water bath.) Explore your ideas; you'll soon find lots of interesting ways to explore Venturi's invention!

Emperor Xianfeng

Emperor Xianfeng and the Chinese Windlass

5

In the summer of 1860, the Second Opium War was reaching its climax. Three years earlier, Britain and France had invaded Canton, China, in order to expand their trade in such disagreeable and corrupting commodities as opium. The Chinese Emperor at the time was Xianfeng, one of the last supremely powerful rulers of China.

Xianfeng did not like what the Western invaders were attempting, so he fought back by sending his soldiers to stop the British from trading in such unsavory goods. But Xianfeng's troops were badly matched against the far better armed Europeans. Unwilling to capitulate to the Westerners, the Chinese embarked on a strategy of low-intensity but exhausting guerilla-style war tactics. The hostilities between the two sides continued in fits and starts for many years. Finally, the European nations grew tired of the expense and anxiety of fighting a war that was so distant, so they launched an invasion fleet that carried nearly 20,000 men to settle matters quickly.

After they landed at the port of Beitang, the soldiers fought their way westward into Peking (now called Beijing) in October 1860. Not long after, the Europeans signed a treaty with the Chinese and an uneasy peace was restored.

The Western world got a bit more out of the deal than just a treaty and trading rights, however. Some British soldiers, encamped on the outskirts of Peking in the aftermath of the final battle, took note of an ingenious device that Xianfeng's engineers used to raise and lower drawbridges throughout the city. The device consisted of a lifting hook that was suspended from an axle made up of cylinders of unequal diameters (see Figure 5.1).

The lifting hook, marked D in Figure 5.1, was connected to cylinders A and B by an enormous length of rope. During operation, this rope spooled off of one cylinder of the windlass, went around the pulley (labeled C in Figure 5.1), and coiled onto the other cylinder. The soldiers observed that the device was capable of lifting huge loads with little effort. The device soon became well-known among Western engineers as the Chinese windlass.

The *Chinese windlass*, also known by its more scientifically descriptive name, the *differential windlass*, is easy to construct and produces enormous mechanical advantage. The lifting power comes from the way

the rope is wound around two cylinders of slightly different diameter. When the handle is turned to lift the load, the rope is payed off the smaller diameter cylinder and on to the larger one. The larger cylinder ends up winding up a bit more rope than is unwound from the smaller one. It is this small difference in the diameter of the cylinders divided by 2 that raises the load with every turn of the handle. So, although raising the load is very slow and requires many turns of the handle, what is lost in speed is gained in power. With cylinders of just a slightly different diameter, even a small person can lift a very heavy load.

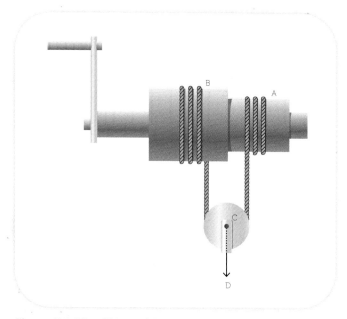

Figure 5.1: The Chinese/differential windlass

Building a Chinese Windlass

If you have access to a lathe, you can machine a high-capacity windlass from a 3-inch-diameter bar mounted on bearings with one end turned to a slightly smaller diameter. But for those of you without a lathe, never fear, you can easily construct a demonstration Chinese windlass model from PVC pipe, pipe fittings, and a few odds and ends commonly found around the workshop or at the hardware store.

Tools

Cordless drill

1″ spade bit

⁷⁄₃₂″ drill bit

⁷⁄₈″ spade bit

¼″ tap

Materials

(1) 3″ PVC pipe, 10″ length

(1) 3″ PVC end cap

(1) 2″ PVC pipe, 10″ length

(1) 2″ PVC end cap

(1) 3″ to 2″ PVC reducing fitting

(1) ¼″ bolt, 2½″ length

(1) ¼″ bolt, 2″ length

(1) Pulley

(1) Lifting hook

(1) ½″ iron pipe, 28″ length, both ends threaded

(2) ½″ pipe elbows

(2) ½″ pipe nipples, 4″ length

(2) ½″ pipe caps

20′ of 1/8″ diameter cord

(2) Pieces of 2″×6″ pine, 24″ length (uprights)

(1) Piece of 2″×6″ pine, 36″ length (base)

(2) Triangular pieces of plywood, 6″ on a side and ½″ thick (gussets)

(6) 2½″-long deck screws

(12) 1½″-long deck screws

PVC primer

PVC cement

Building the Windlass

Follow these steps to make your windlass:

1. Drill a ⅞-inch diameter hole in the center of each end cap (see Figure 5.2).

Figure 5.2: End caps with holes drilled

2. Assemble the windlass axle by cementing the pipes, end caps, and 3-inch to 2-inch coupling as shown in Figure 5.3, using PVC primer and cement. Be sure to read and follow label directions for these items.

3. Drill two ⁷⁄₃₂-inch holes in the axle assembly as shown in Figure 5.3. Take care to drill the holes perpendicularly to the pipe surface. The ¼-inch bolts that go in these holes (in step 11) will act as set screws against the iron pipe inside; it is this contact between the screw ends and the pipe that will allow the windlass assembly to rotate when you turn the crank. Tap each hole with a ¼-inch tap.

4. To assemble the windlass base, drill a 1-inch-diameter hole in each wooden upright, as shown in Figure 5.4.

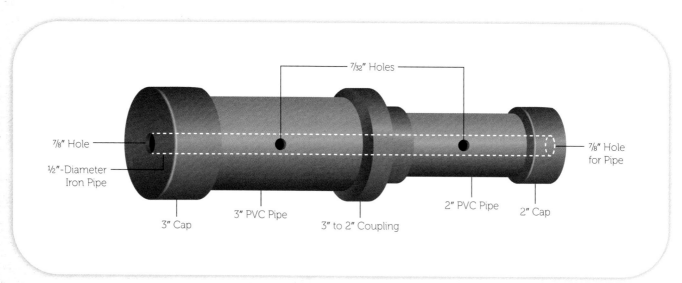

Figure 5.3: Assembling the windlass axle

5. Attach one of the uprights to the base. Use three 2½-inch-long deck screws to fasten the first upright support to the base.

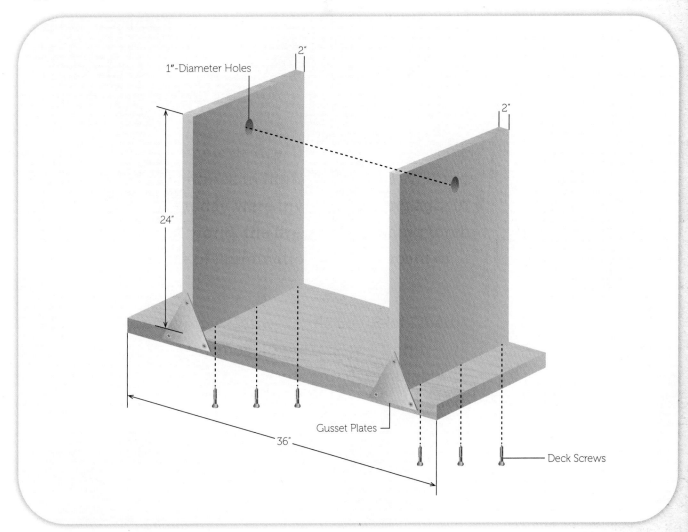

Figure 5.4: Base assembly diagram

6. Insert the iron pipe into the PVC windlass assembly through the holes in each PVC end cap. Center the windlass assembly on the pipe.

7. Insert one end of the pipe into the hole in the upright that is already attached to the base.

8. To attach the second upright, first slide it into place. Then lay the frame on its side and attach the upright to the base using three 2½-inch-long deck screws (refer back to Figure 5.4 as needed).

9. Once you've securely attached the upright to the base, attach the wooden gusset plates for stability to the base and uprights using three 1½-inch long deck screws per plate as shown in Figure 5.5.

Figure 5.5: Attaching the gussets to the base

10. Now that you've got the axle assembly attached to the base, it's time to add the crank. Begin by adding the two iron pipe elbows and the 4-inch pipe nipples to the pipe. Add end caps to the open pipe ends. Your crank should look like the one in Figure 5.6.

Figure 5.6: Windlass with crank added

11. Insert the ¼-inch bolts into the tapped holes on the windlass and tighten them gently until the ends of the bolts make firm contact with the iron pipe (see Figure 5.7).

12. Now wrap the cord around the axle and tie the cord off on the ¼-inch bolts as shown in Figure 5.8.

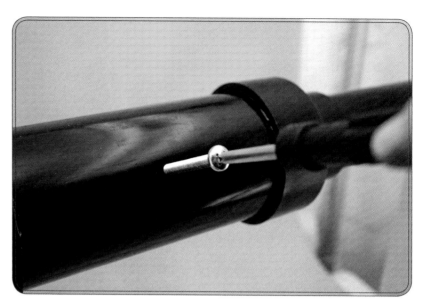

Figure 5.7: Inserting set screws

Figure 5.8: The cord has been attached to the ¼-inch bolts.

13. Place the pulley and hook on the cord as shown in Figure 5.9.

Your Chinese windlass is now ready for use!

Figure 5.9: The finished windlass

Troubleshooting

Here are some pointers for assembling your windlass:

- Be sure the ¼-bolts are seated firmly against the iron pipe; otherwise, the crank will not turn the windlass.

- Make certain you wind the rope as shown in Figure 5.8. If the rope is not arranged correctly on the windlass, the system will not work.

There's No Free Lunch in Physics!

Of course, in the world of physics and engineering, there's no such thing as a free lunch, as you'll soon see.

When a machine like the Chinese windlass requires a small input force—for example, we need to use 10 pounds of force to turn the crank—but produces a lifting force of, say, 100 pounds, we say that the windlass gives us a mechanical advantage of 10 to 1. Lots of machines are designed to yield a big mechanical advantage. For instance, the common automobile jack produces a huge mechanical advantage, and that allows the average motorist to lift two tons of automobile in order to change a tire.

But, there's a hidden cost involved in using a jack, a block and tackle, a Chinese windlass, or any simple machine that allows you to multiply a force. The hidden trade-off is that no machine can increase the magnitude of a force without causing an increase in the distance through which the force must be applied at the same time. Put another way, when a machine produces an increase in force, there is always a proportional decrease in the distance that the object can be moved.

Let's go back to the example of the car jack that allows you to lift a 2000-pound car with only 20 pounds of input force. The trade-off is that you need to move the jack handle 100 inches in order to raise the car a single inch. (Actually, it's more than that since there's friction in the jack mechanism that you must overcome as well.)

Take a look at Figure 5.10 where this is explained for the Chinese windlass.

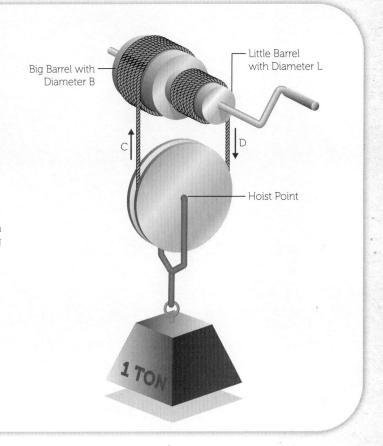

How the Chinese Windlass Works

A Chinese windlass lets even a small person lift a very heavy weight. How is this possible?
Well, the answer is a bit complicated, but if you think about it for a while, you can understand.

When the crank handle is rotated, Cord C winds up onto the big barrel and Cord D unwinds from the little barrel. If you turn the crank one full revolution, a length of rope equal to T1 × diameter B winds up onto the big barrel and a length of rope equal to T1 × diameter L unwinds from the little barrel.

As that happens, the hoist point on the pulley will ascend by half the net difference moved by Cords C and D. So, the amount the weight rises with each turn is (T1 × Diameter B) – (T1 × Diameter L).

If B and L are close in size, the mechanical advantage will be very large: it will take many turns to lift the weight, but the force applied will be very small.

Big Barrel with Diameter B

Little Barrel with Diameter L

C

D

Hoist Point

1 TON

Figure 5.10: The science behind the windlass

Louis Poinsot

Louis Poinsot and the Dancing Spheres

6

Early in the education of every mechanical and civil engineer comes a period of long and detailed study into how forces cause things to move or not move. When you're talking about a car or an airplane, you generally want those forces to make the vehicle move—the general term used to describe this field of study is called *dynamics*. For a bridge, a cell phone tower, or any other object that you don't want to move, you turn to the study of *statics* for understanding.

Any real-world object can have a multitude of forces and loads acting on it, and those forces have technical names such as tension, compression, shear, and torsion. If you also add in twisting forces, called torques and moments, you'll find that designing real-world things becomes complicated quickly.

Luckily, the great minds of the past looked at these difficult problems and figured out how to make the design and engineering of things manageable.

Louis Poinsot, the Father of Geometrical Mechanics

Louis Poinsot was one of those geniuses upon whose work the world of engineering mechanics is founded. Through Poinsot, and scientists like him, we are able to mathematically model, analyze, and understand machines and structures. As a French mathematician of the 19th century, he spent his career in the ivory towers of French academia, formulating the difficult-sounding, yet oh-so-important, field of geometrical mechanics.

Poinsot was the first to demonstrate that any number of individual forces pushing or pulling on a rigid object can be simplified into just a single linear force and a single twisting force known as a *couple*. The great value of the idea, according to Poinsot himself, is that it allows engineers to think of the motion that a large and complex rigid body undergoes in terms that are much easier for the engineer to work with (see Figure 6.1).

Why is this important? Consider the problems facing the engineer who is attempting to design the hull of a sailing ship or the vanes of a windmill. The engineer knows that forces will act on this object from all directions, while friction simultaneously acts on various components, such as the vanes of the windmill and the hull of the ship, as they slice through air and water. All of these forces and torques are of myriad magnitudes and directions. At this point, optimizing the design might seem a nearly intractable problem to the engineer, given the complexity of the interplay of all these forces.

Thanks to Poinsot, however, it is possible to understand what's going on so that we can successfully design boats, masts, buildings, and numerous other things that have many different forces acting on them. Poinsot figured out that all the forces acting on such objects can be manipulated mathematically; instead of dealing with a hundred different quantities, the engineer can use a technique called *vector algebra* to reduce all the forces down to a couple. This simplification was a breakthrough in the field of engineering, and it made the design of all sorts of complicated moving and spinning things possible.

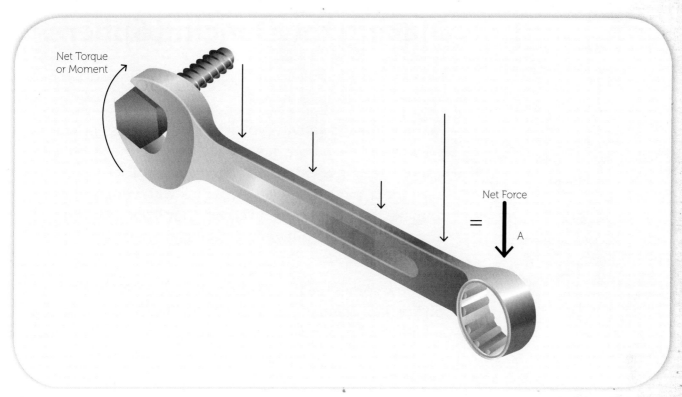

Net Torque or Moment

Net Force

=

A

When you place a wrench on a nut and push the wrench handle, you apply a force. Although the force may be applied all along the length of the wrench, you think of it as being applied only at point A. The arrow at point A is the net, or the mathematical equivalent of all the forces being applied.

The net force at A makes the bolt turn clockwise. The clockwise twist is called a torque, or moment. Poinsot said many different forces and many different moments can be reduced down, mathematically, to a single force and a single moment.

Figure 6.1: Forces and torque

(2) Unhardened steel balls,
5/8" in diameter*

(1) 1/8"-diameter hardwood
dowel, 1/2" in length

(1) 1/4"-diameter flexible rubber or plastic tube, 30" length

(1) BIC Round Stic
ballpoint pen

All-purpose glue

(1)1/2"-thick piece of wood
that is 4"×4"

A wok, pan, or cookie sheet

* Don't use ball bearings; the
heat treatment they've already
received makes them difficult to drill
holes into. A good source for buying
soft steel balls is Craig Ball Sales at
www.craigballsales.com.

Tools

Drill press

1/8" drill bit

1/2" drill bit

Pliers

(2) C-clamps

Making the Dancing Spheres

Here's a fun and novel way to demonstrate Poinsot's forces, torques, and couples at work.

In the following activity, you'll join two small steel balls by drilling a shallow hole in each and connecting them with a short steel or wooden rod. You can then place them on a flat metal pan and give them a flick with your fingers to make them spin. When you do, they'll dance merrily around, darting about while rotating at amazingly fast rotational speeds.

But now comes the interesting part. If you blow a jet of air on the balls, their angular velocity (the speed at which they rotate) reaches almost unbelievable speeds. They spin so fast that the individual balls are not really visible—they look like whirls!

This system of steel balls interacting with the surface on which they spin and the forces acting on them is very complex. But when we apply the work of Poinsot, this can be modeled, analyzed, and understood.

Fabricating Poinsot's Dancing Spheres

The overall goal here is to connect two small steel balls by gluing them to a small axle. The key to success is to drill the holes as perpendicularly to the ball's surface as possible. Don't worry if it takes you a few tries to drill the hole correctly—luckily the balls are quite inexpensive!

1. Begin by securely clamping your 4-inch square piece of scrap wood to the table of your drill press.

2. Insert the ½-inch drill bit into the drill press chuck and drill a hole almost but not quite through the scrap wood (see Figure 6.2).

3. Now, without moving the table, remove the ½-inch drill bit and insert the ⅛-inch drill bit.

4. Place one of the steel balls in the hole you just made in the scrap wood (see Figure 6.3).

5. Start the drill press and slowly press the quill (which is the name of the moving part of the press that holds the spinning drill bit) down to make a ⅛-inch-diameter hole that is ¼-inch deep into the steel ball.

   If the ball begins to spin, carefully hold it in place using pliers. The hole you drilled in the wood should hold the ball in alignment so you obtain a perfectly centered hole, perpendicular to the surface of the ball.

6. Repeat steps 4 and 5 for the other ball.

Figure 6.2: Drilling a hole in the scrap wood

Figure 6.3: Position the steel ball

Figure 6.4: The dowel glued into one ball

7. Place a thin layer of glue on the wooden dowel and insert the dowel into one steel ball (see Figure 6.4).

8. When you're sure the dowel is inserted all the way to the bottom of the drilled hole, insert the dowel's other end into the hole in the other steel ball. Let the glue dry (see Figure 6.5).

Figure 6.5: Letting the glue dry

9. To make the blowpipe for blowing a jet of air at the balls (see exercise introduction), disassemble the BIC Round Stic and remove its tip, which is the plastic piece that holds the ink ball. Insert the tip into one end of the plastic tube until it can't go in any farther (see Figure 6.6).

10. With a motion similar to snapping your fingers, flick the steel ball assembly into the pan, wok, or cookie sheet so it spins as quickly as possible. This action imparts a force and a couple on the assembly, causing the steel spheres to dance across the surface, spinning for 15 to 20 seconds, or possibly more, depending on how good a flicker you are.

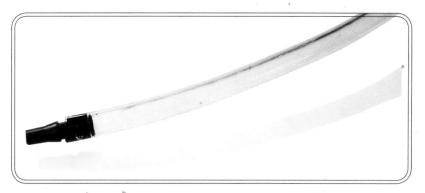

Figure 6.6: Making the blowpipe

11. Make the spheres dance indefinitely by directing a stream of air from your lungs through the blowpipe to one side of the spinning spheres. The jet of air produces a force and a couple as long as you make sure to direct the air at the correct location on the spinning ball assembly. Figure out where to direct the air by trial and error. One you find the right spot, the steel balls spin faster and faster until they become a noisy blur (see Figure 6.7).

Figure 6.7: The spheres of Poinsot in fast rotation

You can vary your experience by shining lights onto your spinning balls to make patterns or by blowing on them with multiple blowpipes from different angles, or by increasing the length of the wood dowel that holds the two balls together.

> **Note:** The steel balls spin with incredible velocity. I recommend you wear safety glasses in case the glue joint fails and the balls come apart.

Simplifying Mechanical Engineering

Although the concepts comprising the study of mechanical and civil engineering have existed since the Egyptians built their first pyramids and chariots, the formalized study of its principles is much more recent.

As the understanding of mathematics and physics replaced rules of thumb and hands-on experience in the 18th, 19th, and 20th centuries, the ability to design structures such as bridges and buildings and understand the actions of complex forces on moving machines became more knowledge-based and therefore more teachable.

For example, imagine a cargo ship being helped to its mooring location by a group of powerful tugboats. There are many forces working simultaneously on the cargo boat: the engines of the tugboats, the wind, the currents in the harbor, and the friction between the ship's hull and the water. But by using the work of Poinsot, the situation can be analyzed and understood quite easily (see Figure 6.8).

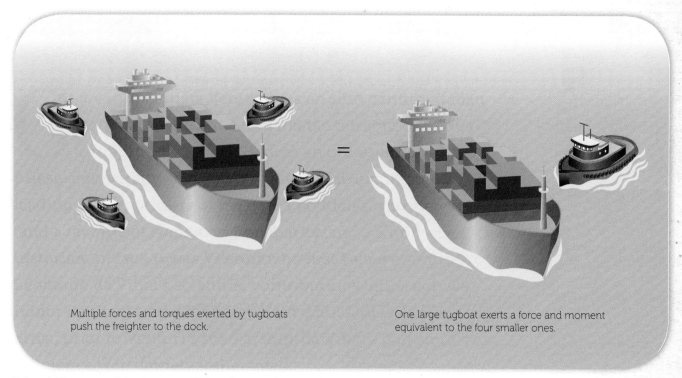

Multiple forces and torques exerted by tugboats push the freighter to the dock.

One large tugboat exerts a force and moment equivalent to the four smaller ones.

Poinsot's Theorem explains that many different forces and torques (torques are twisting forces) can be added together using a mathematical technique called vector algebra.

For example, the ship on the left, above, is being pushed by small tugboats. Each tugboat pushes on the ship in a different spot and that causes the ship to move in a particular direction through the water.

The ship on the right shows how a single bigger tugboat, pushing in exactly the right spot, can have the same effect as the four smaller tugboats. This idea of being able to add forces and torques together to get a single equivalent force and torque is the key to Poinsot's Theorem.

Figure 6.8: Making things simple—going from many forces to one force and one couple

Joseph Louis Gay-Lussac

7

Joseph Louis Gay-Lussac and the Chemistry of Fireproofing

Before homes were electrified, there was a very real danger of clothing catching on fire from the lamps and cooking fires used in daily life. As the study of chemistry became more rigorous, scientists began to think about how to use their knowledge to protect people from this hazard.

The person who first delved systematically into solving this problem was Joseph Louis Gay-Lussac, a 19th-century French polymath. Gay-Lussac was a giant in the history of chemistry, perhaps best known for his work in understanding the properties of gasses.

The Education of a Chemist

Joseph Louis Gay-Lussac was born and raised in the hamlet of Lussac during the turbulent days of the French Revolution. His father spent two years in prison because of his position of power in pre-revolutionary France, but fortunately he managed to keep his head. Joseph was a very bright student and when he turned 19 in 1797, he moved to Paris to study with the eminent scientist Claude Louis Berthollet. Berthollet was a close friend of the recently deceased scientist Antoine Lavoisier, who wrote *The Elements of Chemistry*, the seminal book on chemistry, which first explained fundamental concepts such as the conservation of mass and that chemical compounds were formed from basic elements (see Figure 7.1). Gay-Lussac's interest in chemistry exploded as he studied Lavoisier's textbook, and he showed great aptitude and dexterity for this subject.

Gay-Lussac became known as a skillful experimenter and rose quickly through the ranks of academia to become a professor of physics at the greatest of all French universities, the Sorbonne. By the time he died, he was widely considered to be one of the greatest scientists in France. And although he is perhaps best remembered for his contributions to understanding the nature of gasses, his great scientific work did not end there. His discoveries in the field of inorganic chemistry are of such magnitude

LAVOISIER AND BERTHOLLET
in the Laboratory of the Sorbonne, Paris.

Figure 7.1: Lavoisier and Berthollet

that he might well be considered the father of modern fireproofing.

In his Paris laboratory, Gay-Lussac carried out and documented a host of experiments that formed the basis of the art and science of making a combustible world a bit safer. The key to his success came through his work with one particular substance, the element boron.

If sulfur, often called brimstone, is the element most closely associated with starting fires, then boron is the element most associated with stopping them. Boron is a good fire retardant because it chemically transforms the materials it treats, notably paper and fabric, inhibiting the spread of flame and promoting the formation of a protective layer of char that acts as a fire barrier.

The use of boron was not new in Gay-Lussac's time, as compounds containing the element, such as borax, had been widely used since antiquity. But pure boron was not isolated until 1808. At that time, Gay-Lussac and another scientist, Humphry Davy, one of England's leading scientists (who we meet in the final chapter of this book when we discuss the arc lamp), were locked in a heated competition to isolate boron and thus lay claim to being its discoverer.

To industrial revolutionaries like Gay-Lussac and Davy, the drive to discover new elements was a powerful one. Doing so meant fame and prestige. And isolating and naming an element with properties as useful as boron was particularly compelling.

By 1808, Davy had already discovered and named five elements—barium, calcium, strontium, sodium, and potassium—and he felt confident that he was close to isolating his sixth—the elusive boron. Word reached Gay-Lussac that Davy thought he was close. (In fact, many modern scholars believe that the material Davy isolated that year was indeed boron, but at the time, Davy wasn't able to prove that unequivocally.)

So, across the English Channel, Gay-Lussac redoubled his efforts. Abandoning caution, Gay-Lussac adopted a dangerous laboratory technique involving highly reactive pure potassium metal. By taking that risk, he isolated a substance he called "bore."

Unlike Davy, Gay-Lussac was able to verify to the satisfaction of his peers that what he had found was indeed a new element and therefore he was able to publish his findings.

So, who has the rightful claim as the first to discover boron, Davy or Gay-Lussac?

It depends on how you look at it. Both have their backers, and to many historians, the outcome is too close to call.

In 1821, when Gay-Lussac was experimenting with methods of making materials resistant to fire, he thought back to his work with boron. He saturated fabrics with boron salts and soon found that boron compounds could indeed prevent cloth, paper, and other cellulose-based materials from burning. His success in finding a fireproofing chemical that wouldn't affect the color of cloth or turn it poisonous was a breakthrough. Other scientists built upon his work to make cost-effective and safe-to-use fire-resistant materials for use in theaters and other public spaces.

Fireproofing Paper, Cloth, and Wood

Now that you know a little about the history of fireproofing, let's use Gay-Lussac's discoveries to make a combination fire-resistant hiking stick and campfire poker. This handy, multipurpose item is ideal for overnight hikes into the wilderness. The fireproofing technique we use here is a slight variation of Gay-Lussac's original methods for protecting cellulose-containing materials, but the principles are the same.

Materials that have been treated with a combination of boron-containing chemicals resist burning. That said, it is important to note that this fireproofing method only goes so far. If you leave your stick/poker in the campfire long enough, it will catch on fire.

But, in my experiments, I've found the treatment I outline here is sufficiently effective so that even if the wood stick does ignite, the treatment slows the rate at which flames spread. As always, undertake these projects at your own risk.

Making a Fireproof Walking Stick/ Campfire Poker

Follow these directions to make a hiking stick that will be as useful around the campfire as it will be on the trail.

1. To find a great walking stick, go to a nearby forest and look for a downed branch that is 1 to 1½ inches in diameter, is relatively straight, and is without cracks or knots. Look for a stick that reaches to your sternum, which is a good length.

2. When you've found the perfect stick, remove any small branches that may be growing off the main branch with a hatchet or saw.

3. Then remove the bark with a pocket or similar knife and file off any small protuberances or rough spots. If you want to, you can drill a ¼-inch hole near the top end for a leather or rope loop.

4. Weigh out and then mix the boric acid powder and borax in ½ gallon of hot water in the 1 gallon plastic mixing bucket. Stir vigorously until the chemicals are completely dissolved (see Figure 7.2).

5. Place a piece of cotton cloth in the boric acid solution, saturating it thoroughly.

6. Remove the cloth and hold it over the bucket, allowing the excess solution to drain back into the bucket.

Materials

50 grams of boric acid powder.*

60 grams of borax (available at most grocery stores)

1 gallon of hot water

100% cotton cloth (such as an old undershirt)

1 wooden hiking stick/fire poker

* Boric acid is a weak acid comprised of boron, oxygen, and hydrogen atoms. Although boric acid is deadly to roaches and ants, it is relatively safe for humans to handle and is widely available in hardware stores.

Tools

Hatchet or saw

Knife

File

Drill and ¼" bit (optional)

Metric scale (for measuring grams)

1 gallon plastic mixing bucket

Spoon

Long-handled lighter or fireplace matches

Tall narrow container for soaking stick

Figure 7.2: The mixed boric acid powder, borax, water solution

7. When the cloth is no longer dripping, hang it outside to dry.

8. When the cloth is dry, test its fire resistance by holding a match to a corner of it. The cloth should char and turn black but it should not actually ignite and burn (Figure 7.3).

Figure 7.4 shows the effectiveness of the boron treatment on cotton cloth. You can fine-tune the performance of the fireproofing solution by slightly adjusting the proportions of the borates and water.

Figure 7.3: Testing the cloth

Figure 7.4: The cloth on the right seems to have been soaked in an acceptable solution.

Figure 7.5: Soaking the stick

9. Soak the bottom end of your stick in the solution overnight (see Figure 7.5).

10. Remove the stick from the solution (see Figure 7.6) and let it dry.

11. You now have an all-in-one hiking staff/campfire tending stick. Use the stick to move wood around in the flames of a campfire to improve flame height and fuel use. Remember, your stick is fire resistant, not fireproof, so don't let the tip get too hot or it will still ignite (see Figure 7.7).

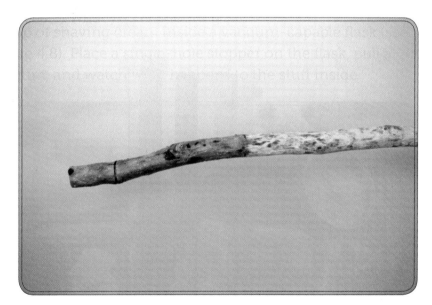

Figure 7.6: The stick after it has been soaked overnight

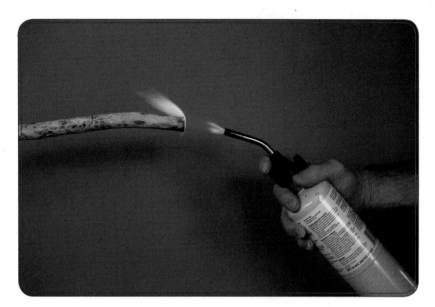

Figure 7.7: If it gets hot enough, your stick will still catch on fire.

To Burn or Not to Burn

How does boron prevent the wood or cloth from igniting? Gay-Lussac found that fire does not occur if air can be prevented from reaching the surface of organic materials by chemically coating the material fibers. Boron compounds are well suited to the task of rendering organic fibers flame-resistant because they cause the normally flammable material to calcinate; that is, when exposed to flame, the fibers turn to char rather than going up in flames.

Now that you know how to make the wood in a hiking stick fire resistant, let's turn our attention to the opposite side of the burning spectrum—how you can make wood, cloth, or paper anti-fire-resistant? That is, how can you turn your cellulose-based object into a form that burns faster, quicker, and more energetically at the merest touch of a flame?

In the stories of Damon Runyon, who wrote about New York City in the days of Prohibition, you'll read about illegal gamblers and their bookies who were always edgy about the police swooping down on them and taking them to jail. When a raid occurred, the police would collect the betting sheets and use them as evidence against the bookies. However, the bookies figured out that if they had a way to get rid of the betting sheets in a hurry—a big hurry—then they wouldn't get convicted.

Eventually, a bookie with some chemistry smarts started using a technique to chemically process the betting sheets in such a way that the smallest flame would vaporize them in a wisp of smoke, leaving only a pile of unreadable ash behind. When the cops broke the door in, all Harry the Horse or Tony Two-Toes needed to do was touch the betting slip with a lit cigar and then he was in the clear.

The chemical process that turns the wood fibers in paper into fast-burning molecules is called *nitration*. This means the cellulose molecules in wood or paper have undergone a chemical process in which they have come into contact with a powerful chemical like nitric acid.

Figure 7.8 shows how treating wood, cotton, or paper with nitric acid changes its molecular structure. When cellulose reacts with the nitric acid, the cellulose molecules acquire many more nitrogen and oxygen atoms on their perimeters. These new "nitrated" molecules are jam-packed with chemical energy and they readily ignite. When nitrated cellulose burns, the chemical reaction occurs with extraordinary speed.

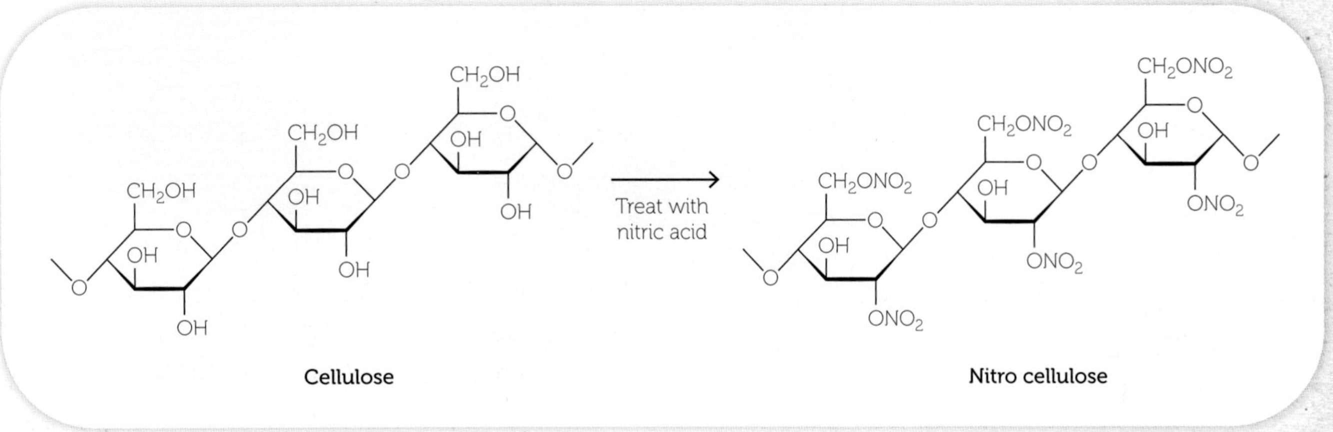

Figure 7.8: The chemical reaction of nitrating

Charles Goodyear

Charles Goodyear and the Vulcanization of Rubber

8

Charles Goodyear may have been the most dogged and unrelenting solo inventor of the 19th century's golden age of invention. But for a fellow with such a famous name, many things about him are misunderstood or little known.

First, Charles Goodyear did not start, work at, or even know of the giant industrial concern called the Goodyear Tire & Rubber Company. The company, which was named in Goodyear's honor, was founded in 1898, about 40 years after Goodyear died.

Second, the incredibly valuable contribution he made wasn't the invention of a thing like a telephone or a light bulb; rather, it was the invention of a process now known as *rubber vulcanization*. And how Goodyear came across it wasn't through his great scientific knowledge or precise experimental technique but by a lucky accident.

The Early Years of Rubber

The story begins 65 years before Goodyear's birth, when, in 1735, the first great international scientific expedition, called the French Geodesic Mission to the Equator, traveled from Paris to Peru in order to accurately measure the shape of the earth. While they were there, the scientists observed the native peoples tapping tropical trees to obtain latex (see Figures 8.1 and 8.2).

The explorers brought some of the stuff back to Europe where it was regarded as interesting, but of little practical use. The one thing they found it could do rather well was remove pencil marks when it was rubbed briskly on paper.

Seventy five years later, though, latex rubber became the rage. In the winter of 1820, a rubber fad swept across America when millions of people bought rubber-coated boots to keep their feet dry. But the fad ended just as abruptly as it started when consumers found that a single summer of hot weather turned their rubber shoes to mush. Nearly all of the rubber manufacturing companies closed. They had discovered that natural rubber clothing just wasn't durable or practical.

Figure 8.1: A tree being tapped

Figure 8.2: Indigenous people playing with a latex ball

It was at this point that Charles Goodyear entered the scene. Goodyear thought that if he could figure out a way to toughen the rubber chemically, he would have a product that people would buy. Although Goodyear knew almost nothing about chemistry, engineering, or business, he was, like the substance he was trying to make, resilient and tough.

Goodyear began to experiment with latex. He mixed in witch hazel, magnesia, and even cream cheese in attempts to turn sticky, soft latex into durable, tough rubber. Nothing worked. He got close a number of times but was not able to develop a repeatable process that would turn rubber into a useful raw material.

He was so determined that he kept trying until he had spent all his money, so he borrowed some. He spent that and borrowed more. Eventually, he and his family were broke and living on the charity of his friends. Things looked bleak.

The Lucky Accident

It was at this point that an accident changed things. On a cold winter day in 1839, Goodyear accidentally brought a piece of rubber that he had treated with sulfur into contact with a hot stove. Afterward, the inventor looked at the piece in amazement. Something incredible had happened. Goodyear wrote

> *I carried on some experiments to ascertain the effect of heat on the (sulfur-treated latex rubber). I was surprised to find that a specimen, being carelessly brought into contact with a hot stove, charred like leather. Nobody but myself thought the charring worthy of notice. However, I directly inferred that if the charring process could be stopped at the right point, it might divest the compound of its stickiness throughout, which would make it better than the native gum. Upon further trials with high temperatures I was convinced that my inference was sound. When I plunged India rubber into melted sulfur at great heats, it was perfectly cured.*

In other words, this rubber fragment did everything that natural latex could not. It had become, in modern parlance, *vulcanized*. It was tough and durable in hot weather, and it stayed flexible in cold weather. This, Goodyear knew, was a product with a huge future.

In the winter of 1841, things were looking up for Goodyear. His new process was an astounding success, and money started to come his way. During his years of experimentation, he had racked up $35,000 in debts (which would be equivalent to about $750,000 today), but he was able to pay off all he owed within a few years.

Unfortunately, Goodyear was not a capable businessman. Although he did obtain a patent for the vulcanization process in 1844, he licensed the process at rates that were far too low for him to make money. Worse, when patent infringers stole his work, he spent more on his attorney's fees than what he was able to recover from the pirates.

He spent the rest of his life attempting to make good on his dream of becoming a millionaire rubber manufacturer. Goodyear staged magnificent displays showcasing rubber products, including furniture, floor coverings, and jewelry, at London and Paris exhibitions in the 1850s. But while he was in France, his competitors were able to circumvent his French patent and his royalties stopped, which left him with outstanding bills he could not pay. Goodyear was thrown into a debtors' prison.

When he died in 1860, Charles Goodyear was $200,000 in debt. But after he died, the royalties on his process started to roll in. His son Charles Jr. later made a fortune manufacturing shoemaking machinery. It's a shame that Charles Sr. never enjoyed the financial success his invention ultimately provided for his family.

Making a Vulcanized Rubber Eraser

In 1770, Edward Nairne, an English engineer, invented the pencil eraser. Prior to this time, the usual way to erase pencil marks from paper was to rub the paper with a piece of bread, which worked fairly well but was rather messy and inconvenient. When Nairne inadvertently picked up a piece of rubber instead of bread, he found that it worked far better. Sensing a commercial opportunity, Nairne began selling rubber erasers, and apparently made a lot of money doing so.

In this activity, you will learn how to make your own erasers.

Making Your Rubber Erasers

Follow these steps to make your erasers:

1. Determine how much Pliatex, water, and vinegar you need to make your erasers by calculating the volume of your mold. Use enough liquid ingredients to fill twice the mold volume for each eraser, because the solidified rubber occupies considerably less space than the liquid ingredients do.

2. Mix equal amounts of water and Pliatex in a bowl or cup and stir the mixture until it is smooth (see Figure 8.3). You can add food coloring to the final product if desired.

Materials

1 pint Pliatex Mold Rubber.*

White vinegar

Water

Food coloring (optional)

* Pliatex is a compounded natural rubber latex product that has been partially vulcanized. It is available at craft stores and online. Internet search terms: pint Pliatex mold rubber.

Tools

Two bowls or cups

Measuring spoon

Stirring spoon

Candy mold

Figure 8.3: Stirring the Pliatex and water solution

3. Measure an amount of vinegar that is equal to the amount of water you used in step 2 and place it in the other bowl or cup.

4. Add the Pliatex-water mixture to the container with vinegar (see Figure 8.4). Stir briefly until the solution congeals into a cheesy, soft mass.

5. Working quickly, move the rubbery mass into the candy mold and press it firmly (see Figure 8.5).

Figure 8.4: Adding the Pliatex mixture to the vinegar

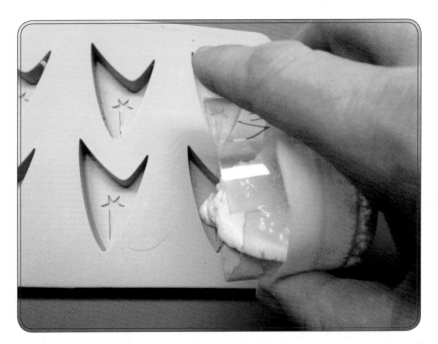

Figure 8.5: Putting the mixture in the candy mold

Figure 8.6: Heating the mixture to make more pliable erasers

6. Pour off the surface water and press firmly again. Continue pouring off any pooling water that appears on the surface of the mold when you press the mixture.

   If you let the rubber harden in the mold, it will turn into a harder, tougher eraser.

7. If you want a softer, more flexible eraser, place the mold and the eraser compound in a 300°F oven for 10 to 15 minutes, depending on the size of the mold (Figure 8.6).

8. Remove your eraser(s) from the oven, remove it from the mold, and let it thoroughly cool and dry (see Figure 8.7).

Your eraser is ready for use!

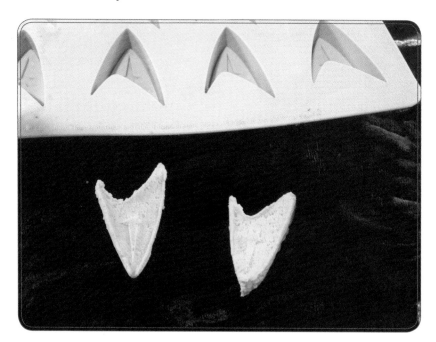

Figure 8.7: The finished drying erasers

The Science Inside: Making Your Eraser

From a scientific standpoint, what's happening is this: Pliatex is a partially vulcanized latex rubber compound called polyisoprene. It will naturally begin to coagulate and harden on its own unless it is stored in an alkaline environment. The manufacturer adds ammonia to keep the pH level high during storage. When you take the cap off the bottle, you'll smell the ammonia used to preserve it.

When you add vinegar to the solution, the acetic acid in the vinegar quickly lowers the pH, and the large polymer molecules in the latex come out of solution to form a solid piece of rubber.

Although natural latex is soft and sticky, vulcanized rubber is tough and resilient. Our pencil eraser, which is made from partially vulcanized rubber, sits halfway between the two extremes. The eraser removes mistakes from a written note because the rubber is hard enough to rub out the graphite particles from the paper, yet it is soft enough to not tear the paper.

Figure 8.8 further explains the differences between natural and vulcanized rubber.

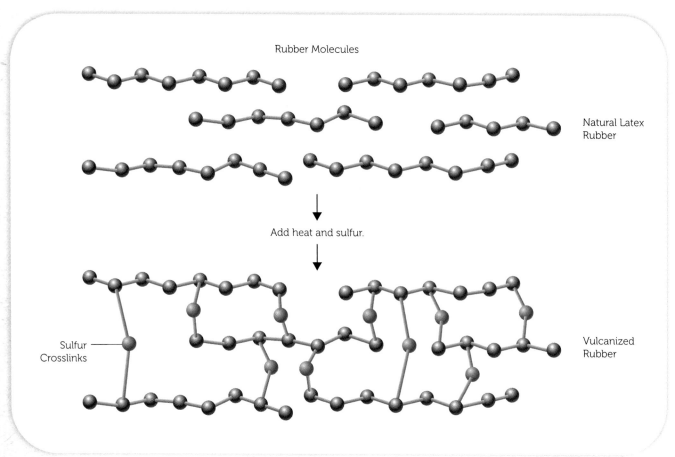

Natural latex rubber is made up of long chains of molecules, called polymers, in long strips. The chains are easily pulled apart, so natural rubber is fairly weak and not durable.

Charles Goodyear figured out that when you heat latex while adding sulfur, chemical crosslinks form between the latex polymer chains. Vulcanized rubber is ten times stronger than natural rubber, but it is still elastic. Great stuff indeed!

Figure 8.8: The differences between natural and vulcanized rubber

Part III

Electricity for the Future

In the final three projects, we'll make and use humankind's most industrious servant, electricity. Electricity is a bit hard to understand because you can't see it or hold it in your hand. But, in order to really learn about electricity, we need to figure out how to get our arms around it, at least figuratively. As you already are aware, you can't really do that with household current—there's far too much energy in there to play safely. Consequently, the amount of electricity we'll use in our projects will stay very small; but we'll still have enough to work with to do some very useful and interesting things.

Electricity was known to the early Greeks; indeed, the word *electricity* derives its name from the ancient Greek word for amber which was "electron." For generations, Greek philosophers credited amber with nearly magical powers.

When the Greeks rubbed amber with a cotton cloth, they said that "the spirit hidden inside" came out. The spirit could grab and hold on to small detached objects, such as bits of parchment, thread, or cloth. The ancients were at a loss to explain this. Perhaps the gods up on Olympus were behind it. In any event, the phenomenon was looked upon with superstitious awe and the amber itself was regarded as having magical attributes.

As time went on, it was discovered that this mysterious attractive power could be found in a lot of places, not just with amber. You could simply rub a piece of glass rod with silk or leather and generate electricity. So, electricity, it was reasoned, was not a property of amber or glass or silk but came from a far more basic, and presumably, earthly source.

The type of electricity produced by rubbing or friction was impossible to control so it wasn't of much use. But in the early 19th century, scientists found that simply pressing two different materials together could induce a constant flow of electricity. With this advancement, electricity had real value. It could produce continuous light, plate metal, and energize magnets.

By the end of the century, electricity was being produced in large quantities in the spinning components of generators and dynamos. Whole cities became electrified and the world changed faster than it had ever changed before.

In the following projects, we will see that electricity is electricity, whether it's produced by generators in power plants, by static friction, via thermo-electricity, or by chemical batteries, and it can be harnessed and utilized to do wonderful things.

Benjamin Franklin

Benjamin Franklin · Explores Electricity

9

There are two stories that just about every American elementary school student knows: the one about George Washington chopping down the cherry tree and the one about Benjamin Franklin flying his kite in a thunderstorm to prove the electrical nature of lightning bolts. Both are good stories, but they differ in one very important respect: Franklin actually *did* fly his kite in the thunderstorm, whereas the story about the cherry tree is a myth.

Few people have made as large an impact on daily life as Benjamin Franklin. Aside from his political and literary contributions, he was colonial America's greatest scientist. It seems that nearly half the words we commonly use to describe our daily interactions with electricity—positive, negative, charge, discharge, battery, cell, conductor, condenser—were coined by Franklin.

The kite story was first related by the famous English chemist Joseph Priestley in his 1767 book, *The History and Present State of Electricity*. Priestley may have embellished the event a bit, but even if some of the details are hazy, the overall import is unquestionable. In his legendary experiment with the key, the kite, and the thunderstorm, Franklin proved what many educated people suspected—lightning is indeed electrical in nature.

The story goes something like this. On June 10, 1752, the skies darkened as clouds rolled into Philadelphia from the west. Keeping an eye on the gathering storm, Franklin and his son Billy headed to an empty field on the north side of town with their kite. They put the kite aloft as the wind picked up. The wind began to blow in earnest, and soon the kite, framed from cedar twigs and covered with silk cloth, strained against the kite string. Franklin and his son waited inside a shed, no doubt impatiently, until the storm began. When he looked down, Franklin noticed that the kite string's thin strands of hemp were moving. Seemingly of their own volition, they stood as erect as soldiers at attention.

Attached to the kite string was a metal key. Nearly shaking with excitement, Franklin turned his hand, moving his knuckle toward the key. Suddenly, he felt a small shock as the electricity jumped from the storm-electrified key to the grounded Franklin.

The storm continued, wetting the kite string, and conducting increasing amounts of electricity down into the small shed where Franklin collected the charge in a primitive type of capacitor called a Leyden jar. Figure 9.1 further clarifies the science going on in Franklin's experiment.

In retrospect, Franklin's method of studying electricity was extremely dangerous. Of course, we know better than to attempt such an experiment today, but in Franklin's time, he was navigating very new areas of scientific knowledge.

In his famous experiment, Franklin attached one end of a silk cord to a kite and the other end to a metal key. Then he ran a wire from the key to a Leyden jar, which is a simple electricity storage cell. Finally, he attached a silk ribbon to the key as well. Franklin flew the kite in a thunderstorm, allowing the cord to get wet but keeping the ribbon dry. Negative charges ran down the wet cord to the key and into the Leyden jar. Franklin did not get electrocuted because the dry ribbon insulated him.

Figure 9.1: Explaining Franklin's dangerous experiment

Science versus Superstition

Franklin turned his discovery of the electrical characteristics of lightning to great practical use; he built on this knowledge to invent the lightning rod. The lightning rod has been a tremendous lifesaver since it was invented. When installed correctly, it was effective in preventing the lightning-caused fires that frequently destroyed churches and other tall buildings.

But old ways die hard. Even after Franklin explained the true nature of lightning, people continued to attribute it to supernatural origins. In fact, many religious leaders continued to think lightning was supernatural and ascribed it to the work of demons. To defeat these demonic hordes, they ordered that church bells be rung during thunderstorms. We now know that this was not just silly, it was downright dangerous. In the years following Franklin's experiment, more than 120 bell ringers were electrocuted when lightning ran down the wet bell cords into their bodies. They were dead ringers indeed!

The Leyden Jar's History

The Leyden jar that Franklin used to store the trapped lightning was a simple capacitor that consisted of two conductors separated by a thin insulating sheet. This device was discovered accidentally in Leyden, Holland, just a few years before Franklin's experiment. A Dutch scientist named Pieter van Musschenbroek was attempting to electrify a container filled with water by touching it with an electrified glass bottle. By chance, the design of the container included two conductors separated by a thin insulator. When he accidentally touched two parts of the container simultaneously, the Dutchman found himself stunned, literally and figuratively. News of this incredible "Jar of Leyden" spread like wildfire throughout the scientific community. At last, here was a way to hold and store electricity.

Improvements to the original design of the Leyden jar were made as experimenters tried different techniques, and finally the device now commonly used in experiments and presented in the activity section that follows came to be.

For a while, it seemed as if everyone had a Leyden jar and people all over 18th-century Europe enjoyed shocking themselves and one another for fun and profit. Itinerant lecturers called "electricians" traveled from town to town, demonstrating the wonders of the recently discovered ways to control and utilize the phenomenon now called electricity. Perhaps the greatest electrician was Jean-Antoine Nollet (also known as Abbé Nollet), a member of French King Louis XV's court. Nollet once arranged for 180 people to hold hands in a great room in the king's palace. He then cranked up his electrostatic generator (not unlike the electricity generating machine you build later in this chapter) and asked the man at the head of the line to touch a brass ball. Upon contact, all 180 people jumped into the air, mightily surprised by the shock they received from a big Leyden jar. The king, wrote Abbé Nollet, was delighted.

You can simply re-create Franklin's experiments without the incredibly dangerous activity of flying a kite in a thunderstorm. The static electricity–generating machine detailed in this activity uses the rubbing friction between two electrically dissimilar materials to generate oppositely charged particles. A "comb" harvests the particles and directs them to a Leyden jar for storage. Once you've got bottled lightning, the opportunities for science and entertainment, as Ben Franklin and Abbé Nollet proved, are endless.

Materials

- (1) 2″×10″ board, 2′ in length
- (2) 2″-diameter PVC pipes, 16″ in length (uprights)
- (2) 2″-diameter PVC tee fittings, cut as shown in Figure 9.2 (upright supports)
- (4) 2″-long lag screws and washers
- (1) ½″-diameter PVC pipe, 20″ in length (roller support)
- (2) ½″-diameter PVC pipes, 5″long (crank pieces)
- (2) 1/2″-diameter elbow fittings (crank fittings)
- (1) ½″-diameter cap fitting (crank fitting)
- (1) 3″-diameter PVC pipe, 10″ in length (roller)
- (2) 3″PVC cap fittings (roller caps)
- (1) ¾″outside diameter (OD) copper pipe
- (1) ¾″copper pipe cap fitting
- (1) Round brass knob with bolt
- (8) #10 brass screws, 1½″ in length, or aluminum flashing cut into a comb with eight or more points
- Fur
- Glue

Making Your Own Static Electricity Generator

The electricity produced by frictional machines (or simply by rubbing your stocking feet over carpet on a dry winter day) is extremely high voltage, but of extremely low amperage. Getting a static-electricity shock from a frictional machine is much like being hit by an extremely small water balloon dropped from a very great height; you sure feel it, but it doesn't last long.

Except for some laboratory applications, static electricity isn't of much practical use. One exception is electrostatic cleaning equipment, which uses high-voltage, low-amperage wire grids to filter dust and other small particles out of the air.

Making and charging a Leyden jar is a fun and interesting project. If you make one, you are following directly in Benjamin Franklin's footsteps. Take care not to get shocked!

> **Note** The dimensions I've provided here are just guidelines. You can make your generator just about any size you want.

Making the Generator

Before you begin constructing your generator, please refer to Figure 9.2 to see the relationship of the parts.

Tools:

Drill

Hack saw

Screwdriver

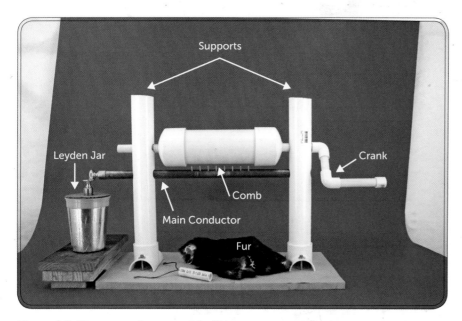

Figure 9.2: The generator assembly diagram

Now it's time to begin assembling your electrical experiment:

1. Cut the bottom section off the 2-inch tee fittings so they lie flat on the base and attach them to the frame with lag screws and washers, as shown in Figure 9.3.

2. Drill ¾-inch holes in the middle of each PVC cap so the roller support piece can be inserted as shown in Figure 9.4.

Figure 9.3: Modified tee fittings

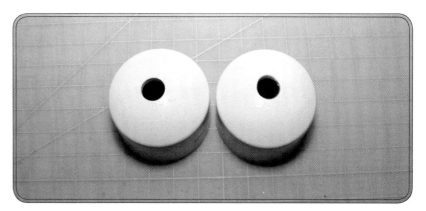

Figure 9.4: Drill holes in the caps.

3. In each 2-inch PVC upright, drill a ¾-inch hole; these holes are also for the roller support.

4. Drill the ⅞-inch holes in the 2-inch PVC uprights 2 inches below the ¾-inch holes (as shown in Figure 9.5).These are for the copper pipe.

5. Now attach caps to the 3-inch pipe and insert the roller support through the holes in caps, as shown in Figure 9.6.

6. Drill parallel ⅛-inch-diameter holes at equal intervals in the copper pipe.

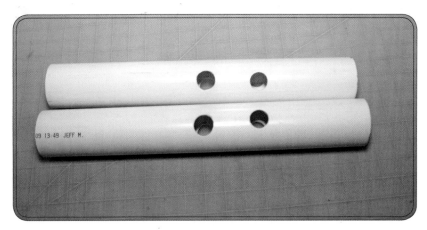

Figure 9.5: PVC uprights with holes drilled

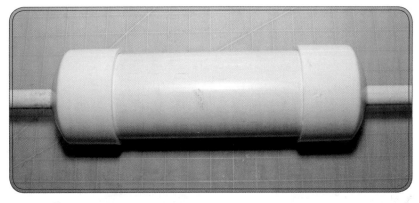

Figure 9.6: The roller assembly

7. Insert the brass screws into the holes (see Figure 9.7).

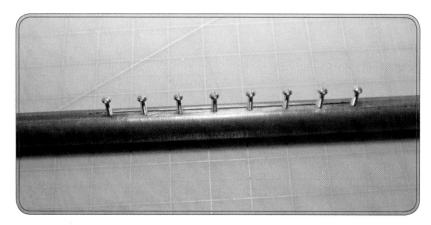

Figure 9.7: The charge collector

8. Attach the copper cap to the copper pipe and affix the brass knob to the end.

9. Now insert the copper pipe and roller support pieces into holes in the uprights (see Figure 9.8).

10. Once you got this assembled, place the uprights into the upright supports and glue them into place with PVC cement according to the directions on the cement bottle. Adjust the brass screws so that the screw heads lightly brush up against the PVC roller assembly.

11. Attach the two ½-inch-diameter, 5-inch-long pipes, the two ½-inch-diameter elbow fittings, and the ½-inch-diameter cap to the end of the ½-inch PVC pipe extending through the roller assembly in order to make a crank. When you've got the crank attached, it should look like Figure 9.9.

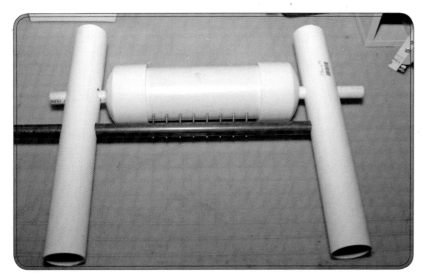

Figure 9.8: PVC and copper pipes placed into uprights

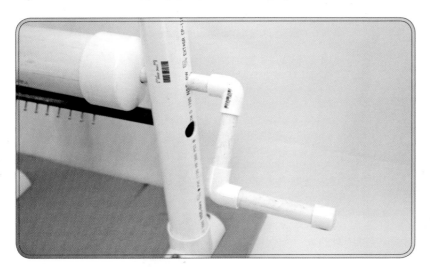

Figure 9.9: Crank handle detail

12. To get your assembly ready to generate some electricity, adjust the screws on the copper comb so the screw tips just brush against the 3-inch PVC pipe when you turn the crank. (Refer back to Figure 9.8 for a visual.)

Building the Leyden Jar

Materials

Plastic jar

Aluminum foil

Brass knob

Copper or brass beaded lamp chain

Now that you've built a static electricity generator, you'll need to build a device to store that charge until you're ready to use it. You'll now construct a simple capacitor called a Leyden jar to hold the electricity you generate.

To assemble your Leyden jar, follow these steps (refer to Figure 9.10 for a visual guide):

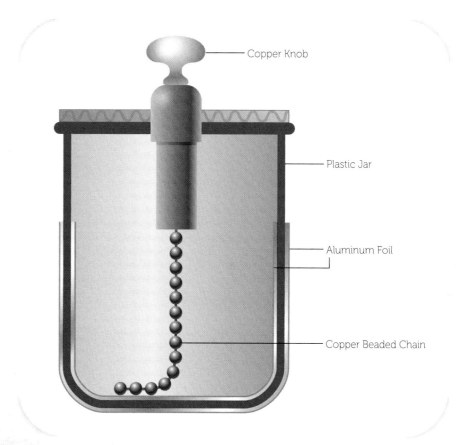

Figure 9.10: The Leyden jar assembly diagram

1. Place a piece of aluminum foil on the inside bottom and lower inside walls of the jar.

2. Place a piece of aluminum foil on the outside bottom and lower outside walls of the jar.

3. Drill a hole the same diameter as your brass knob in the plastic top of the jar. Insert the knob and glue it into place if necessary.

4. Attach the beaded chain to the knob and allow the other end of the chain to lie upon the aluminum foil on the inside bottom of the jar.

Using the Leyden Jar

By turning the handle of the static friction machine, you can generate a sizable static electricity charge and store it within the confines of the Leyden jar until you're ready to discharge it.

1. Place the Leyden jar knob so it rests touching the knob on the copper pipe in your generator assembly as shown in Figure 9.2 earlier in this chapter.

2. To generate electricity, press the fur firmly against the 3-inch PVC pipe and rotate the crank rapidly.

As you turn the crank, the copper pipe becomes electrified and the charge is collected in the Leyden jar. If you touch the knob of the charged Leyden jar with your finger, you will induce a spark. Large sparks are painful, so you may want to discharge the Leyden jar by using a piece of curved wire so that you can touch the knob and the outside foil simultaneously. Use a thick rubber glove or something simi-lar to hold the wire.

Remember, Leyden jars can pack an electrical wallop! Don't leave charged jars lying around—if you do, somebody may get a nasty surprise. Also, don't shock people who don't want to be shocked, and limit the size of your jar; a quart-sized jar is more than enough.

This project works far better on dry winter days than on damp summer ones. The less humidity in the air, the better.

From Leyden Jars to Hydroelectric Dams

Although our modern understanding of electricity began with Franklin and his experiments, electricity proved to be such an interesting phenomenon that many other great minds quickly decided to follow in Franklin's footsteps. Soon after the kite and Leyden jar experiments, Alessandro Volta and other scientists in Europe began experimenting with the chemical production of electricity by means of batteries.

You really can't overstate the importance of batteries to the progress of electrical invention because they provided a way to make "current" electricity, which flowed continuously, instead of popping and then almost instantly disappearing like the spark from a Leyden jar.

With continuous-current electricity, useful inventions such as the telegraph and telephone became possible. And not long after that, one of England's greatest scientists, Michael Faraday, discovered electromagnetic induction. In his London laboratory, Faraday found that a varying magnetic field makes electricity flow in an electric circuit. This led to the invention of the dynamo, which produces direct current, and ultimately, to the electrical generator, in which motion from the wind, flowing water, and expanding steam can produce alternating current electricity in great quantities.

Alessandro Volta

Alessandro Volta and Electrodeposition

10

If you're like most people, you'd be hard-pressed to name more than a few living scientists or engineers. But 200 years ago, scientists were the rock stars of the day, and those at the top of their profession were more famous than nearly any entertainer or politician of their time.

For example, consider the Italian scientist Alessandro Volta, who invented the first electrical battery. If batteries seem a bit ho-hum to you, you'd best reconsider. When he announced his discovery, the excitement in the scientific community was unlike anything seen previously. Not only did he become rich and famous, but Napoleon Bonaparte was so stirred by Volta's battery that he not only conferred upon him the Legion of Honor award, but he also made Volta a French count.

Why was such honor heaped upon Volta? Because from nearly the moment it was invented, Volta's battery was a world changer. Before the battery, the only way to produce electricity on demand was with a static-electricity—making friction machine and a Leyden jar. Except for their appreciation of a few parlor tricks associated with the Leyden jar, the society of the time didn't have much practical use for it. The problem was that the electricity generated from the friction machine and the Leyden jar produced a single, big, instantaneous spark that wasn't usable in any practical way.

Volta Harnesses Electricity

In early 1800, Volta set to work with a great deal of enthusiasm and dedication, for he had an idea that he believed was not just new, but extraordinarily useful as well. How right he was!

Working long hours in his laboratory at the University of Pavia in Italy, he carefully investigated his idea that, under certain conditions, dissimilar metals placed next to each other could generate an electrical voltage. If this was so, he reckoned, then scientists could produce constant, even flows of electricity. And if that was possible, electricity would no longer be simply a curious scientific phenomenon, but instead, it would be a useful technology that could lead to a brighter future for everyone.

In the laboratory, Volta stacked alternating plates of zinc and copper into a tall column similar to the way they appear in Figure 10.1. Between each group of plates he inserted squares of cloth that he had moistened with a weak saline solution. This was the first "battery," meaning a device that produced a steady source of electrical current.

All over Europe, the excitement over Volta's electric battery invention was palpable. Sir Humphry Davy called the voltaic battery "an alarm bell to experimenters in every part of Europe." Soon, scientists were applying constant current electricity to just about everything they could, eagerly studying and documenting the effects.

Less than six weeks after Volta wrote to the president of London's Royal Society with the results of his findings, two British chemists, William Nicholson and Anthony Carlisle, were able to decompose water into its two

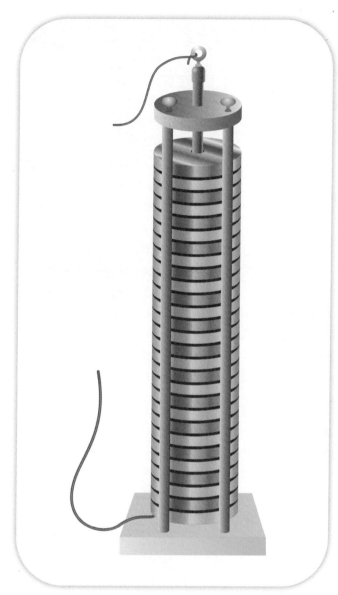

Figure 10.1: A voltaic pile

component gases by applying electricity from a battery based on Volta's design.

This discovery was soon followed by Davy's production of electric light from a continuous spark in the gap between electrodes in a carbon arc light (see Chapter 11, "Humphry Davy and the Arc Light"). These were important discoveries indeed, and they were soon followed by one of most commercially important of all early electrochemical applications: electroplating.

Before Volta and the other chemists of the early 19th century teased out the basics of electrochemistry, metal items were basically made from one solid hunk of metal; they were the same on the inside, the outside, and everywhere in between. For example, in those days, silver forks and knives were 100 percent silver, which made them very expensive. But using Volta's battery, small amounts of metal salts, and the right catalysts, the Italian scientist Luigi Brugnatelli discovered that it was possible to treat or "plate" inexpensive metals with a thin, hard, and economical coating of pure nickel, copper, silver, or even gold. Factories were soon mass-producing immense quantities of silver-plated teapots, silverware, hairbrushes, snuff boxes, and more, making life a little more refined, even luxurious, for the working classes. It is this electroplating technique that you'll become familiar with in this chapter's activity.

Making a Copper-Plated Sign

Materials

Copper sulfate*

Sulfuric acid**

Dishwashing liquid

A copper sheet, cut into 1"×4" strips (Found in the metals display rack in hardware stores)

A brass sheet, cut into 1"× 4" strips (Or any size that works well for your design)

Glass bowl or pan—1½ quart or larger

(2) Copper wires—16 gauge, 4" long

1'×4' wooden board, long enough to span the glass bowl with (2) ¼"-diameter holes drilled (as shown in Figure 10.2)

1.5-volt D-cell battery and battery holder, or low-voltage DC power supply

(2) Alligator clip patch cords, 18" long

Wax candle or nail polish

* Typically found in its pentahydrate form in the drain-cleaner section of hardware stores. Manufacturers include Rooto and Roebic.

** Also often found in hardware stores for use as a drain cleaner.

Tools

Small (2 to 6 oz.) measuring cup

32 oz. measuring cup

Glass stirring rod

Drill with a ³⁄₁₆" bit

Rubber gloves

Splash-proof eye protection

In this project, you'll use Volta's battery technology to make a sign by electrically plating copper onto a piece of brass using materials you can find in most hardware stores. As I mentioned earlier, there are many types of electroplating, and professionals can apply platings of gold, silver, chromium, nickel, and other metals. Most of these metals require the use of dangerous chemicals such as cyanide and arsenic. However, as long as you approach this project with care, it's easy and fun to plate copper on brass without needing to use such extremely dangerous chemicals.

Now that you've assembled all the materials you will need to make this project, it's time to get started. Before you do, take a look at Figure 10.2, which shows the setup you'll need for plating copper onto a piece of brass in order to make a sign.

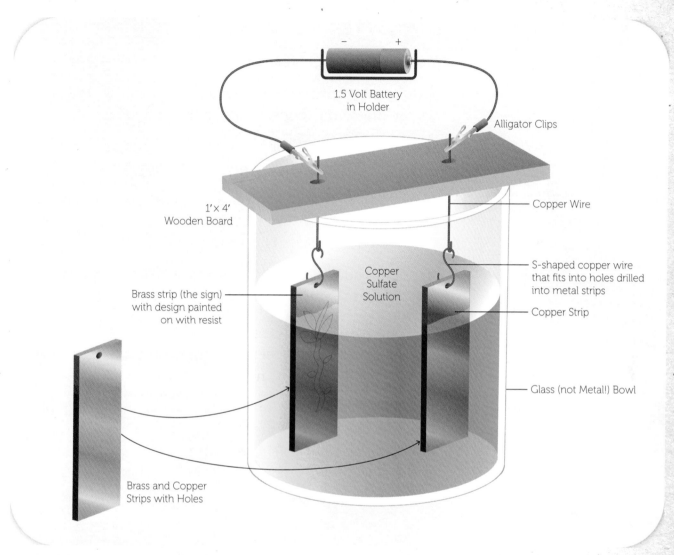

Figure 10.2: Electroplating process diagram

Part 1: Prepare the Materials

First you need to get ready by preparing your materials.

1. Begin by drilling ³⁄₁₆-inch holes in the middle top part of the brass and copper strips as shown in Figure 10.2.

2. Add 6 ounces of copper sulfate to 24 ounces of water in the large measuring cup. Stir until dissolved.

3. Put on your rubber gloves and splash-proof eye protection for the next several steps and for all steps thereafter that require you to handle the acid solutions.

4. Slowly add 3 ounces of concentrated sulfuric acid to the copper sulfate solution (see Figure 10.3). Be careful with this stuff!

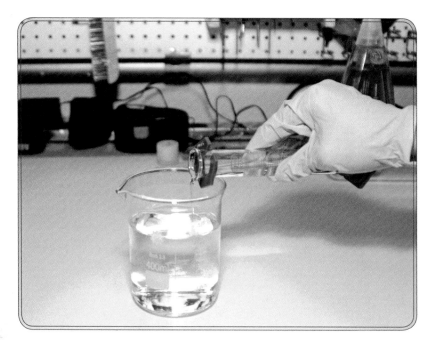

Figure 10.3: Adding concentrated sulfuric acid

5. Pour the solution into the large glass bowl.

6. Rinse the large measuring cup thoroughly.

7. Now pour 12 ounces of water into the large measuring cup. Slowly and carefully add an ounce of sulfuric acid to the water in the measuring cup. Stir with the glass stirring rod.

8. Thoroughly clean the brass and copper strips with dish-washing detergent and water. Rinse them well.

9. Use the nail polish or candle wax to draw a design on the brass strip (see Figure 10.4). Technically, the wax or polish is called a *resist*. The design you draw with your resist will be the only part of the brass strip not plated, so you are actually making a negative image.

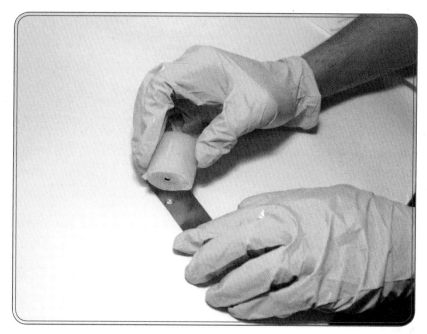

Figure 10.4: Using candle wax to make a resist

Part 2: Pickle the Metal

Now that you've prepared your metal strips, it's time to make sure your strips are clean for the electroplating process:

1. While you're waiting for your resist to dry, bend the ends of the 16-gauge copper wire so they form hooks.

    Refer back to Figure 10.2 for the following portion of this project.

2. After your wax or nail polish design is dry, hang the hooks you just made over the rim of the measuring cup so the copper and brass strips are immersed in the diluted sulfuric acid.

    The acid bath removes impurities, such as stains, contaminants, and rust or scale from the metal surface so the plated metal applies evenly. Electroplating professionals refer to this operation as *pickling*.

3. Remove the strips after 2 minutes.

4. Pour the diluted acid down the drain.

Part 3: Electroplate Your Sign

Now, you're ready to begin plating:

1. Place the wood board so it spans the rim of the electroplating bowl (as shown in Figure 10.2).

2. Insert the copper wire hooks through the holes in the board and attach the alligator clips to the non-hooked ends of the wire so that the hooks below will hang

submerged in the copper sulfate solution when you've attached the strips.

3. Hang the copper and brass strips from the hooks. Make sure the side of the brass strip with the resist design is facing the copper strip.

4. Attach the free end of the alligator clip patch cord that is connected to the copper plate to the positive battery terminal (as shown in Figure 10.2).

5. Attach the free end of the remaining alligator clip patch cord that is connected to the brass plate to the negative battery terminal.

6. Let the electricity flow for approximately 2 minutes.

7. When 2 minutes is up, remove the strip with the design and rinse it with lots of water.

8. Rub the electroplated strip with a soft cloth. Your electroplated sign is complete (see Figure 10.5)!

Figure 10.5: The completed sign

Safety Notes

As you probably figured out when I told you to put on gloves and safety glasses, the chemicals involved in this experiment can be dangerous if not handled appropriately. Sulfuric acid is an especially powerful corrosive and you must not let even the tiniest drop contact your skin.

Here are a few things to remember:

1. "Do like you oughta, add acid to water."

In steps 4 and 7 of Part 1 of this project (and any future occasions that involve the use of acids) it's important that you always pour acids into water rather than vice versa. Pouring water into an acid causes the mixture to boil and splash and, of course, that's the last thing you want to occur.

2. "Little Johnny took a drink, but he will drink no more, For what he thought was H_2O, was H_2SO_4."

   Stuff like sulfuric acid (H_2SO_4) must be handled carefully. Always be aware of what you're doing.

3. Wear the protective gear listed and take care to avoid spills.

4. Dispose of the leftover solutions carefully and responsibly. You can flush small quantities of dissolved copper sulfate and sulfuric acid down municipal drains, but do so only in small quantities and use plenty of water.

Electroplating Explained

Soon after Volta invented his battery of electrical cells, he showed his invention to his good friend, Luigi Brugnatelli. Although Brugnatelli was a physician, he was intensely interested in his friend's invention. He began to experiment with the voltaic cell, and in 1805, he discovered that he could, if the conditions were just right, make electrically charged metal particles travel from one electrically charged pole to another. Brugnatelli figured out that by applying electricity to a solution containing metal salts, he could cause a thin coating of metal to form on one of the poles.

This discovery became the basis for the process of electro-plating. Figure 10.6 shows what is happening during the electroplating process.

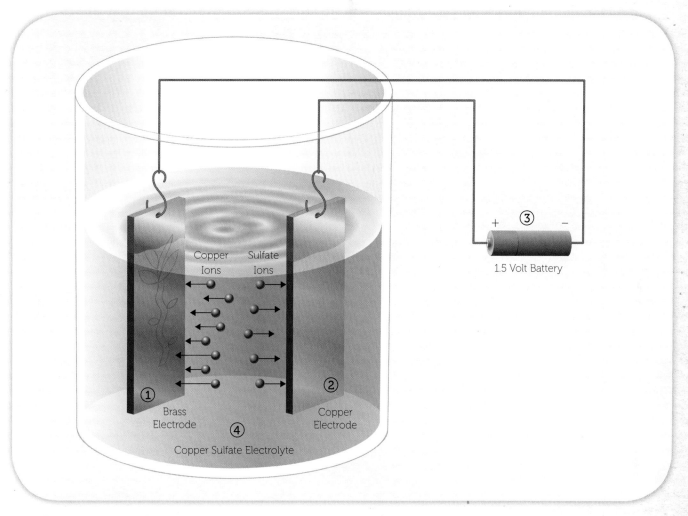

Copper Ions

Sulfate Ions

① Brass Electrode

② Copper Electrode

④ Copper Sulfate Electrolyte

+ ③ −

1.5 Volt Battery

When the copper and brass electrodes are submerged in the solution and a voltage is applied, current flows.

The brass strip (1) becomes the negative electrode, or anode, and the copper one becomes the positive electrode or cathode (2). With the voltage applied (3), the copper sulfate solution (4) splits apart, producing copper ions and sulfate ions. The copper ions are positively charged and are deposited on the brass strip, except where the resist is applied. The sulfate ions move to the copper strip simultaneously, causing the copper strip to dissolve. The ions move to the cathode, plating it.

Figure 10.6: Atoms on the move

Humphry Davy

Humphry Davy and the Arc Light

11

Despite what you've heard, Thomas Edison did not invent the first electric light. Although there is still some controversy about this issue, it does seem that he may deserve credit for the first electric light bulb. But more than 70 years before Edison's 1879 incandescent lamp patent, the English scientist Humphry Davy developed a technique for producing controlled light from electricity.

Sir Humphry Davy was one of the giants of 19th-century science. A fellow of the prestigious Royal Society, Davy is credited with first isolating elemental sodium, strontium, calcium, and barium and with inventing the miner's safety lamp. The safety lamp alone was directly responsible for saving hundreds if not thousands of miners' lives.

But it is Davy's invention of the arc light for which we remember him here. In a sense, it was Davy's light that lit the Industrial Revolution.

Davy and the Arc Light

Davy's artificial electric light was actually a controllable electric arc. The light it produced emanated from an assembly that was comprised of carbon rods made from wood charcoal connected to the terminals of an enormous collection of voltaic cells. In Davy's day, thousands of voltaic cells (which were invented by Alessandro Volta, who we met in Chapter 10), similar to modern chemical batteries, had to be wired together in a series to produce the voltage required to strike an arc between the carbon electrodes.

When Davy closed the switch that connected the voltaic cells to the electrodes, electricity jumped between the carbon tips. The resulting glaring lucent dot of white heat was so bright that it was dangerous to look at for more than a split second.

"The Dazzling Splendor," as Davy called it, was a tricky beast to control, however. After the initial sparks appeared between the electrode tips, Davy had to separate the carbons slightly and carefully in order to sustain the continuous and bright arc of electricity. Once he had accomplished that, he found the device could sustain the arc for long periods of time, even as the carbon rods were consumed in the heat of the process.

But Davy's arc light was not economically practical until the cost of producing a 50-volt power supply became reasonable. This didn't occur until mid-1870s with the introduction and commercialization of the electrical dynamo. But as soon as they became affordable, arc lights were everywhere, archetypically in searchlights, as well as in lighthouses, in street lights, on movie sets, and in movie projectors.

It took a lot of juice to run a searchlight. To maintain its arc, a 60-inch-diameter, World War II vintage, carbon arc searchlight drew about 150 amps at 78 volts, which is equivalent to a 12,500 watt light bulb. A lot of power? Yes, but it could light up an airplane from miles away.

Perhaps the largest carbon arc light ever made was the 80-inch-diameter monster searchlight built at the turn of the 20th century by General Electric. It lit the grounds of the 1904 St. Louis World's Fair with its billion candlepower arc light.

Making a Davy Carbon Arc Light

The light from an arc light results from a spark or electric arc jumping through the air between two carbon rods. Now, it's a rather finicky sort of light source as the gap between the rods must be of just the right size. If the gap is too big, then the arc will not form, and if the gap is too narrow, then it will produce less light.

But despite these drawbacks, arc lights were better than gas lights. By the late 1870s, electric generators were capable of providing large amounts of electricity dependably and cheaply. Electric arc street lighting was installed in many cities, including London, Paris, and Los Angeles.

The project that follows uses the high-amperage, low-voltage power from a battery or battery charger to strike and maintain a bright arc between two pure graphite rods.

The easiest way to obtain pure carbon rods is to cut open a regular *non*-alkaline AA, C, or D cell battery with a Dremel tool or a hacksaw. Such batteries are usually labeled "heavy duty." Cut off the top of the battery and carefully remove the carbon rod from the black, greasy packing that surrounds it (see Figure 11.1). The packing material will stain hands, clothes, and work surfaces, so wear rubber gloves and cover surfaces with newspaper.

Materials

12V (volt) battery charger with an ammeter or an 18V battery (from, say, a portable power drill)

(2) Carbon rods, ½" in length by approximately ¼" in diameter

Tools

Dremel style rotary tool or hacksaw

Wire stripper/wire crimper

Needle-nose pliers (for bending wire)

Sandpaper

Screwdriver

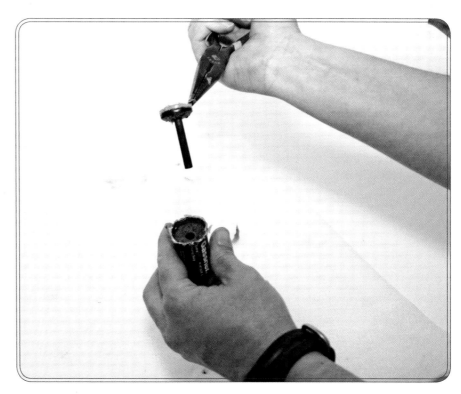

Figure 11.1: Removing the carbon rod

Building an Arc Light

Our carbon arc light produces light due to the presence of an electric arc spanning the air gap between two electrodes. Take care to avoid touching the wires and other components because they become very hot when the electric current is flowing.

First take a look at Figure 11.2 and then follow these steps to build your arc light:

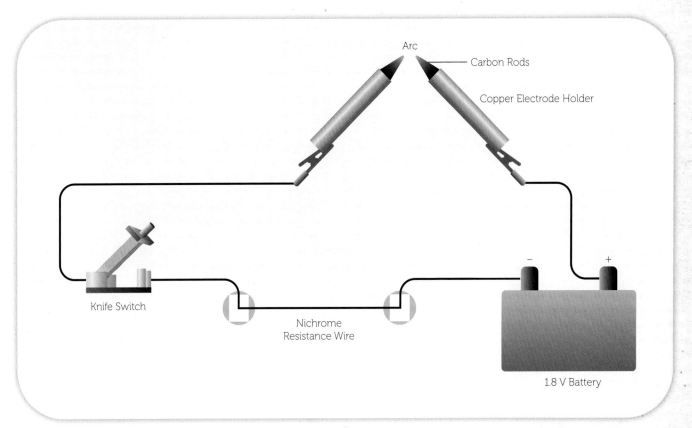

Figure 11.2: The assembly diagram

1. Carefully clean the carbon rods you obtained from the non-alkaline batteries.

2. Sand the carbon rods into electrodes—that is, remove material with sandpaper until they fit snuggly into the ¼-inch-diameter copper tubes.

3. Crimp the carbon rods in place into each of the two copper electrode holders. Sand the protruding end into a point as shown in Figure 11.3.

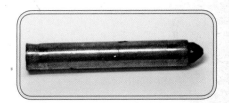

Figure 11.3: Electrode installed in the electrode holder

4. Now mount the posable clips to a wood frame. Position the electrodes so the carbon points are almost touching, as shown in Figure 11.4.

Figure 11.4: Electrodes and holders installed on the box

5. Mount the on/off switch as shown in Figure 11.5.

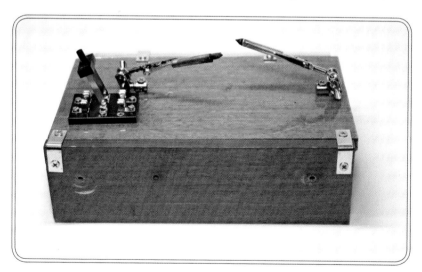

Figure 11.5: Adding the on/off switch

6. Next, mount the ceramic insulators as shown in Figure 11.6. You should mount the insulators approximately 10 inches apart.

Figure 11.6: Installing the insulators

7. You're now ready to wire the circuit as shown in Figure 11.2.

The circuit is very simple. Electricity from the battery goes to the first electrode holder, through the carbon electrode, and across a very small spark gap to the second electrode. From there, the electricity goes through the main on/off switch and then across a length of nichrome wire before it enters the opposite pole of the battery or battery charge.

8. Close the switch and carefully adjust the spark gap until you obtain a bright white light as shown in Figure 11.7.

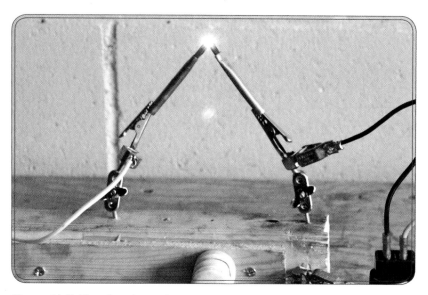

Figure 11.7: The electric arc in use

9. Once a bright arc is struck and maintained, you can optimize the light output of the system by making the nichrome resistor wire longer or shorter.

Notes: Adjusting Your Arc Light for Best Performance

Here are some things you need to keep in mind during the process of making your arc light:

1. Every homemade arc light is a bit different. You'll need to make position adjustments as necessary.

2. The spacing of the electrode gap is critical. Take your time adjusting it in order to obtain the best arc light. Too much or too little contact will result in no arc light (see Figure 11.8).

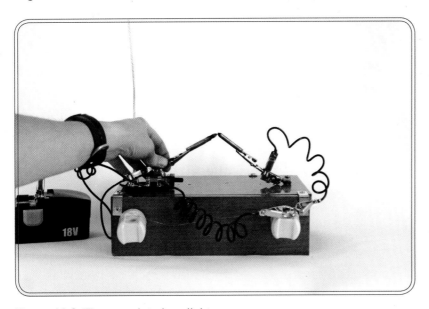

Figure 11.8: The completed arc light

3. Your battery will be damaged if the circuit is run without an adequate load. The nichrome wire provides just enough resistance to prevent the battery from shorting. You will have to adjust the length of the nichrome wire for best performance. If the wire length is too short, it will quickly burn up. If the length is too long, the arc light will be dim. Only trial and error will help you determine the correct nichrome wire length.

Using Your Arc Light Safety

In order to make sure you stay safe during the process of making and using an arc light, keep these tips in mind:

1. If made correctly, the arc of an arc light is *very* bright and emits UV light. Never look directly at it.

2. The nichrome wire and copper electrode holders get extremely hot. Be very careful around them or you are apt to get burned! The good news is there is little to no shock hazard associated with 12 and 18 volt DC batteries.

3. This light is a demonstration device only and you should only operate it intermittently and for brief intervals. Running the arc light for too long can damage your battery or battery charger. If you are using a battery charger, check the ammeter on your charger to make sure the circuit is not shorted. If it is, or nearly is, use a longer nichrome wire.

How an Arc Light Works

Although making an arc light isn't terribly complicated, the arc light's underlying physical processes are indeed complex. Although it is normally a non-conductor, carbon will conduct electricity in certain circumstances. The graphite rods used in arc lights conduct electricity, albeit grudgingly, if enough electrical potential is applied. Figure 11.9 visually explain show an electric arm forms.

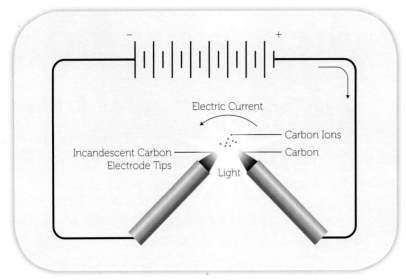

Sir Humphry Davy's carbon arc light consists of two pieces of carbon whose ends are separated by a small distance. When an electric current is applied, the carbon electrode tips vaporize due to the heat. The vapor contains carbon ions, which are good conductors of electricity. As electrons flow through the gap, the electrode ends become incandescent and radiate light. At the same time, the heated carbon vapor in the arc gap also produces visible light.

Figure 11.9: The arc light explained

Appendix: Going Further, Deeper, and Higher

After becoming acquainted with the work of the great inventors, engineers, and makers of the Industrial Revolution, perhaps you'd like to read more about their projects. Well, there are plenty of resources available! Here are a few ideas:

1. Visit *www.makezine.com*.

   This is the Internet home of *Make:* magazine, which is a bimonthly magazine that provides a ton of do-it-yourself (DIY) ideas in each issue. The projects involve computers, electronics, robotics, metal-working, woodworking, and other disciplines.

2. Visit a Makerspace in your area.

   A Makerspace is an area devoted to making DIY projects. It's a place where people come together to share ideas, tools, and expertise. Visit *http://spaces.makerspace.com/makerspace-directory* to see if there is one near you.

3. Many books are available with additional ideas for exploring science, history, and DIY. My books, *Backyard Ballistics*, *Defending Your Castle*, and *The Art of the Catapult*, are full of interesting projects. Find them and others online or in your local bookstore.

The Inventor's Workshop

If you enjoy making things, eventually you'll need a space to work and tools to work with. In the days of the early inventors, a tool was simply a handheld implement, such as a hammer, saw, or file, and it was used for performing or facilitating mechanical operations, like cutting, pounding, or filing. But in modern times, tools do so much more. They measure quantities and qualities precisely, they join electrical components into circuits, and they perform a hundred other useful operations.

The Workbench

First and foremost, you'll need some sort of flat, solid surface on which to work. Any sturdy table will do, but a workbench is a great help, because it provides the foundation you need in order to work skillfully.

You can make or buy a workbench. Many lumberyards sell prebuilt workbenches or kits containing all the materials you'll need. You can also find a design for one or draw one yourself. Designs for homebuilt workbenches run from complex Scandinavian designs with beechwood frames that are mounted on self-leveling hydraulic cylinders down to a simple plywood door nailed to two sawhorses. No matter what sort of bench you have, the addition of a wood vise and pullout shelf make it more versatile.

Necessary Tools

Ask an expert what sort of tools to buy and the typical advice is to buy the best quality tools you can afford. In most cases, that's good advice. Cheap screwdrivers, for example, can be a big mistake; the soft metal edges of inferior blades can bend or even break under stress, and the plastic handles chip when you drop them. For any tool you use frequently, it makes sense to go with quality.

On the other hand, when you've got a one-off job, and you're not sure if you'll ever have another use for piston ring pliers or a gantry crane, buying an inexpensive tool may make sense.

Here are some ideas for outfitting your workspace.

Basic Tools

These handheld tools are useful in a wide variety of situations. They are as important for adjusting or repairing existing items as they are for making new ones.

Screwdrivers. Choose an assortment of good-quality Phillips-head and flat-head (and, possibly, Torx) screwdrivers in a variety of sizes.

Handsaw. Most often, you'll be cutting dimensional lumber (2′× 4′s, 2′× 6′s, etc.) to size, so choose a saw with cross-cut instead of ripping teeth.

Hacksaw. You need this type of saw for those occasions when you have to cut through something harder than wood.

Hammers. Start with a claw hammer for nailing and a rubber mallet for knocking things apart.

Socket and wrench set. If you work on things mechanical, you'll appreciate the quality of a good socket set. Spend the money and get English and metric sockets as well as Allen wrenches.

Pliers. Pliers come in a variety of shapes. At a minimum, your shop should have standard, needle-nose, and vise grips.

Cutters and mat. You'll want diagonal cutters, a utility knife, tin snips, a wire cutter/crimper/stripper, and a good pair of scissors. You'll also find a self-healing cutting mat to be a great help. Buy one at any fabric store.

Clamps. Clamps securely hold workpieces, allowing you to work safely and accurately. Clamps come in various sizes and are selected based on the size of the workpiece.

Linear measuring tools. Make sure you have a tape measure, a protractor, and a combination square.

Files and brushes. You'll need flat and round bastard files and a wire brush. (A bastard file refers to a file that has an intermediate tooth size.)

Mixing and volume measuring equipment. Stock your work area with plastic bowls in different sizes, disposable spoons, measuring cups, and measuring spoons.

Safety equipment. Safety glasses, hearing protection, a fire extinguisher, goggles, a dust mask, and gloves are all very important. All safety glasses, even inexpensive ones, must conform to government regulations, so they all provide adequate protection. However, more expensive ones are more comfortable and look better, making you more inclined to always use them.

Cordless and/or corded electrical drill. A drill with a variety of screwdriver tips and drill bits may well be your most frequently used power tool. Corded drills are lighter and more powerful, but many people appreciate the flexibility of a cordless model. The larger the top-end voltage (e.g., 14.4 or 18 volts) of a cordless drill, the greater its torque and the more it weighs.

Specialty Tools

Inventors often need specialized tools to perform certain tasks. They are typically not expensive, at least for entry-level tools.

Soldering iron. Choose a variable-temperature model with changeable tips.

Magnifying lens. You'll find a swing-arm magnifier with a light to be a very helpful addition to your shop. It mounts directly to your workbench and swings out of the way when not in use. It's great for everything from threading needles to examining surface finishes.

Scale. A triple-beam balance or an electronic scale is a necessity for chemistry projects and mixing stuff.

Digital multimeter. If you do any electronics work, a volt-ohm meter with several types of probes and clips is indispensable.

Power Tools

These are great, if you can justify their cost:

Drill press. A sturdy drill press provides far more accuracy and drilling power than a hand drill.

Belt sander. Belt sanders utilize a rotating abrasive belt to quickly remove material from workpieces.

Grinder. Grinders have rapidly spinning abrasive wheels and are used for shaping metal and sharpening tools.

Table saw/Band saw/scroll saw. Electrically powered saws cut wood much faster than handsaws. However, they must be used with great care.

Beyond these basics, there are hundreds, if not thousands, of tools available—all of which may be useful, depending on the project. In regard to stationary power tools, it's a tough call. Because they are expensive and require a lot of shop real estate, it really depends on what you're going to do *most*. I use my table saw all the time, but I know people who consider a band saw to be an absolute necessity, and others who say a scroll saw is their number one power saw priority.

Supplies

Besides raw materials and tools, stock your shop with key general supplies. Here's my suggested checklist:

- Duct tape
- Electrical tape
- Transparent adhesive tape
- Powdered graphite lubricant
- Rope or cord
- String or twine
- Light all-purpose oil
- White glue
- Superglue (cyanocrylate)
- Quick-set epoxy
- Extended-set epoxy
- Sandpaper: fine, medium, and coarse
- Heat-shrink tubing
- Zip ties
- Pencils
- Ink markers
- Rags, wipes, and towels

It takes time and money to accumulate a good supply of tools. But a well-stocked workshop or tool box and the ability to use the tools properly are valuable assets for any inventor.

Further Reading

Do you want to read more about early scientists and technology, or find more do-it-yourself (DIY) projects that are rooted in the timeframe of the Industrial Revolution? These books are excellent places to start:

The Ancient Engineers by L. Sprague de Camp (Doubleday, 1963)

Although his name might not be so well-known now, L. Sprague de Camp was one of the best-known science fiction writers of the 1930s and 1940s. But it's this 1963 work of science fact that I highly recommend. As Isaac Asimov said of it, "it's the history (of science) as it should be told."

The Art of Construction by Mario Salvadori (Chicago Review Press, 1990)

One of my all-time favorites, this book is a fine introduction to the hows and whys of civil engineering. Structures like buildings and bridges stand firm in the face of the forces of wind, earthquakes, and gravity, and *The Art of Construction* explains how that happens.

Atoms under the Floorboards by Chris Woodford (Bloomsbury Sigma, 2015)

This interesting science book asks and then answers scientific questions that are so basic that most people haven't thought to ask them. Why do some things stick together and others don't? Why can you see through glass, but not metal? The answers are as good as the questions.

Make: Fire—The Art and Science of Working with Propane by Tim Deagan (Maker Media, Inc., 2016)

If the amount of fire associated with Abe Lincoln's campaign torches isn't enough for you, this book may provide more of what you seek. Although this book assumes a certain level of maturity and skill with tools, there are few better resources available for learning to work with fire through the building of several interesting propane-related projects.

Backyard Ballistics by William Gurstelle (Chicago Review Press, 2012)

My first book and it's still very popular! This book combines science, history, and the do-it-yourself ethos. Since it was first published in 2001, the dozen or so projects it contains have launched not only a lot of flying projectiles, but have interested a lot of folks in exploring science and technology.

Ancient Inventions by Peter James and Nick Thorpe (Ballantine, 1995)

As we've seen in this *ReMaking History* book, our modern era has no monopoly on clever and important inventions. Cranes, compasses, and clocks were invented long ago, and although most of the names of the inventors of these items are lost in the mists of history, their importance continues. This book sheds light on a great many more.

Index